DHA:

(DOCOSAHEXAENOIC ACID)

THE MAGNIFICENT MARINE OIL

A Health Learning Handbook
By Beth M. Ley-Jacobs, Ph.D.

BL Publications
Temecula, CA

BL Publications, Temecula, CA 92592
(909) 694-6283 Order number: (888) 367-3432

This book is not intended as medical advice. Its intention is solely informational and educational. It is wise to consult your doctor for any illness or medical condition.

Library of Congress Catalog Card Number: 98-093858

Ley-Jacobs, Beth M. 1964-
/ by Beth M. Ley-Jacobs
-- 1st Ed.
p. 120 cm.
Includes biographical references and index
ISBN **1-890766-01-1** (alk. paper)

Printed in the United States of America

First edition, January 1999

ISBN: 1-890766-01-1

Credits:
Research and Technical Assistance:
Sid Shastri, Anthony Almada
Proofreading: Carrie Vom Steeg
Cover artwork:
Roderick Sutterby, Middleside Ninebanks Hexham, Northumberland NE47 8DL England
e-mail: sutterby@globalnet.co.uk

Table of Contents

Introduction 5
Essential Fatty Acids: The Good Fats 7
EPA and DHA 14
DHA: The Brain, Nervous Tissue and Vision 22
Prostaglandins 25
Infant Nutrition Needs 33
Attention-Deficit Hyperactivity Disorder 40
Addiction and other Behavior Disorders 43
Depression 48
Schizophrenia 51
Alzheimer's Disease/Memory 55
Inflammatory Disorders 58
- Allergies/Asthma 62
- Skin Disorders .65
- Arthritis 68
- Crohn's Disease/Ulcerative Colitis 72

Coronary Artery Disease 73
Diabetes 80
- Neuropathy 82

Using Fat for Weight Control 84
Immune System 91
Cancer 94
Salmon Recipes 101
Glossary 108
Bibliography 110
Index 117
About the Author/Cover Artist 119
Other Books from BL Publications 120

My dear friend, I pray that you may be in health; I know that it is well with your soul. **3 John: 2**

Many know the story about how Jesus fed the multitude of people with five loaves of bread and two fish (Matt.14:17), but another important "fish" story is in John 21.

This was the third time Jesus showed himself to the disciples after he was resurrected. They had been fishing all night at the Sea of Tiberias and had caught no fish and had no food for breakfast. Jesus appeared to them on the shore and asked, *"Children, have you any food?"* They answered him *"no."* Jesus then said to them, *"Cast the net on the right side of the boat, and you will find some."* They did, and they were unable to draw the net up because it was so full of fish. They brought the net into shore containing 153 large fish in total. Jesus said to them, *"Come and eat breakfast."* Jesus then came and took the bread and gave it to them, and likewise the fish. The story finishes as after breakfast Jesus asks Simon Peter three times, *"Do you love me?"* Simon Peter responded all three times, *"Yes, you know I do."* Jesus says, *"Feed my lambs."* The second time, Jesus says, *"Tend my sheep."* The final time Jesus responded, *"Feed my sheep. When you were younger, you girded yourself and walked where you wished, but when you are old, you will stretch out your hands, and another will gird you and carry you where you do not wish to go. Follow me."* (NKJV)

Introduction

Health Learning Handbooks are designed to provide interesting and useful information to improve one's health and well-being through what the body needs to obtain and maintain good health.

Good health should not be thought of as the absence of disease. We should avoid this negative disease-orientated thinking and concentrate on what we must to do to remain healthy. Health results from supplying what is essential to the body on a daily basis, while disease results from living without what the body needs. We are responsible for our own health and should take control of it before disease does.

Our health depends on education. Education is power.

Good fats, bad fats, good cholesterol, bad cholesterol. The subject of fats is not so simple anymore. At one time I believed that the best way to health was to eat as little fat as possible. With so many food products now available as "fat free" one could easily be convinced that this is true, but it's not.

The truth is that some types of fat are an essential part of our diet and are very important as structural components in our bodies, the brain and eyes in particular. It is so important that adequate levels are necessary to ward off a very long list of common health problems. These include Alzheimer's disease and other memory disorders, heart disease, diabetes, hyperlipidemia, cancer, inflammatory conditions such as allergies, asthma, arthritis, dermatitis, psoriasis, depression, and some very interesting behavior disorders including ADD/ADHD in adults and children, violence, aggression and addiction disorders such as alcoholism. One of the most surprising effects is weight problems. It was this effect that inspired further investigation which uncovered some pretty interesting facts, and thus the reason for another book.

DHA is not new, we just haven't known much about it other than the fact that it accompanied the fatty acid EPA in cold-water fish. Over the past two decades we have learned that DHA is the critical long-chain Omega-3 fatty acid found in the brain. Research in the areas of schizophrenia, ADD and other neural-related disorders is actually just beginning.

Don't confuse DHA with DHEA, the subject of one of my earlier books, *DHEA: Unlocking the Secrets to the Fountain of Youth.* DHEA stands for Dehydroepiandrosterone, a steroid hormone produced in the adrenal glands.

Also, I have included a glossary in the back of the book so you can easily look up some of the important terms used in the book.

Essential Fatty Acids: The Good Fats

The body can manufacture certain fatty acids from the carbon, hydrogen and oxygen atoms provided by food. Certain fatty acids are non-essential because it is not required that we eat them as they can be produced in the body from essential fatty acids (EFAs) or through other means. We tend to eat fats (mostly saturated) in excess, because they taste good and give texture to many foods, making them more appealing - crunchy, flaky, creamy, etc. They are stable and can be used in cooking and baking.

On the other hand, essential fatty acids cannot be manufactured by the body and must be obtained from the food we eat. They are more difficult to consume because they are highly unstable. They must be consumed fresh and unrefined or they quickly go rancid. Extracted from their whole natural form, they have a very short shelf life. They cannot be used in cooking as they are sensitive to heat, light and oxygen.

There are two main groups of EFAs: the Omega-3 oils (including alpha linolenic acid - ALA) and the Omega-6 oils (linoleic acid - LA). They are both long-chain polyunsaturated fatty acids, meaning they have two or more double bonds between any two carbons in their chain.

If there are adequate quantities available, the body can usually convert ALA into the longer-chain fatty acids, eicosapentaenoic acid (EPA) and docosahexaenoic acid (DHA) through enzymatic conversion with delta-6 desaturase), elongation, and delta-5 desaturation. See chart on page 29 for further clarification.

These desaturase enzymes occur inside many different cellular tissues such as the liver, intestinal mucosa, brain and retina. The activity of delta-6 desaturase is altered by a variety of hormonal and dietary factors. For example, insulin and EFA-deficient diets increase desaturase activity. Glucose, epinephrine (one of the stress hormones) and glucagon (from a high sugar diet) decrease desaturase activity. Genetic factors and other reasons can also interfere with this enzyme disallowing production of EPA or DHA from LNA.

ALA, EPA and DHA each have their own various metabolic activities in the body and cannot be substituted for each other for all of these activities, Therefore, it is important to obtain adequate quantities of each of them.

In the Omega-6 family, LA (Linoleic acid) has 18 carbons, two double bonds, written18:2n-6, converts into Gamma-linoleic acid (GLA),18:3-n-6, which converts into Dihomo-gamma-linolenic acid (DGLA), 20:3n-6. Finally, this can be easily converted into the longer-chain Arachidonic acid (AA), 20:3n-6, especially when there is excess present in the body. This is largely due to the overconsumption of animal products. These polyunsaturated fatty acids form the membranes of every cell in the body and produce biologically active compounds called eicosanoids which influence every process in the cell. These include prostaglandins, thromboxanes, prostacyclins, and leukotrienes which are important for several cellular functions such as platelet aggregability (clotting), inflammation, vasodilation, growth of smooth muscle cells, white blood cells, etc.

As Omega-3 fats have fewer hydrogen molecules compared to Omega-6 fats, Omega-3s are more liquid at room temperature. Northern plants and fish, in response to cold weather, produce more Omega-3 oils than Omega-6. The Omega-3s help keep cell membranes fluid, preventing them from freezing. Southern plants and fish produce fewer Omega-3 oils and more Omega-6 oils.

Most vegetable oils (if consumed fresh) such as corn and safflower, contain some Omega-6 fatty acids, but have very little Omega-3s. For optimal health, both oils must be consumed in the right proportion.

Excessive amounts of Omega-6 may produce effects

which are detrimental to health. Dietary Omega-3s act in two ways to diminish the negative effects of excess intake of Omega-6s. They compete for enzymes involved in elongation and desaturation and by forming eicosanoids (such an anti-inflammatory prostaglandins) which compete at cellular receptors to reduce the effects of Omega-6 eicosanoids (such as inflammatory prostaglandins).

FATTY ACID	**FOOD SOURCE**
Omega-3s	
LNA *(Alpha-Linolenic Acid)*	Vegetable Oils, Nuts, Cold-Water Fish
EPA *(Eicosapentaenoic Acid)*	Cold-Water Fish, Microalgae
DHA *(Docosahexaenoic Acid)*	Cold-Water Fish, Microalgae
Omega-6s	
LA (Linoleic Acid)	Vegetable Oils
GLA* *(Gamma Linolenic Acid)*	Borage Oil, Black Current Oil
DGLA* *(Dihomo Gamma Linolenic Acid)*	
AA *(Arachidonic Acid)**	Meats, Animal Products
Omega 9s	
OA *(Oleic Acid)**	Vegetable Oils

* non-essential in adults

SOURCES OF ESSENTIAL FATTY ACIDS:

	% of Fat Content	% of fat of Omega-3	% of fat of Omega-6
Seed oils			
Canola (rape)	30	7	30
Chia	30	30	40
Corn	4	-	59
Evening Primrose	17	-	81
Flax	35	58	14
Hemp	35	20	60
Olive	20	-	8
Peanut	48	-	29

(cont.)	% of Fat Content	% of fat of Omega-3	% of fat of Omega-6
Pumpkin	47	0-15	42-57
Rice Bran	10	1	35
Safflower	60	-	75
Soybean	17.7	7	50
Sunflower	48	-	65
Walnut	60	5	51
Wheat germ	11	5	50
Fish			
Eel	12-18	18.9- 31.4	1.6-3.6
Mackerel	10	18.9- 31.4	1.6-3.6
Salmon	10-15	18.9- 31.4	1.6-3.6
Sardines	19	18.9- 31.4	1.6-3.6
Trout	10	18.9- 31.4	1.6-3.6
Albacore Tuna	6	25	.8
Halibut	2	25	25
Pike	1-4	N/A	N/A
Carp	1-4	N/A	N/A
Other Sources			
Eggs (hard boiled)			
Supermarket	23	.2	11.3
Flaxmeal fed	26	.8	16

How to Get the Most Value From Your Oils

Omega-3 and Omega-6 oils are fragile and easily destroyed by exposure to light, air, and heat. Natural oils will keep four to six months if properly stored. Safflower, sunflower, and corn oils (polyunsaturates) become rancid more quickly than olive oil (a monounsaturate). Store oils in a tightly capped, dark bottle in your refrigerator to protect from light, heat and oxygen.

Adding vitamin E immediately after opening can help reduce rancidity. Put a hole in the end of a capsule and squeeze the oil into your container, put the cover back on and gently shake a few times. Purchase oils in smaller bottles to insure freshness.

Oils must be unrefined and pressed without solvents. They must be consumed fresh and should taste like the seed they were pressed from. Most oils purchased at the grocery store in clear bottles are highly processed with high temperatures and solvents destroying the oils. These are not acceptable sources of EFAs. The exception is "extra virgin" olive oil stored in a dark container.

Some unrefined oils stored in opaque bottles (not clear) purchased at health food stores may be acceptable sources of essential fatty acids. Oils should say "Fresh" and should have an expiration or bottling date. Not all oils which claim to be cold-pressed are, so "buyer beware."

Avoid Hydrogenated Oils

Avoid altered fats such as unsaturated fats hydrogenized into saturated ones. The process creates abnormal or unnatural fatty acids, called trans-fatty acids. Not only is the body unable to use them as EFAs, but these abnormal fats can block the utilization of the needed EFAs in the body.

Manufacturers use hydrogenation because it prolongs the shelf life of their products. They do not consider the fact that the nutrients lost during the hydrogenation process are the very ones that are most necessary to protect the oils and for the body to best use the oils.

Agricultural methods have also decreased our intake of EFAs. Caged chickens and their eggs, as well as feedlot-raised cattle, are producing much lower levels of Omega-3 and Omega-6 fatty acids than free-range chickens. This also applies to fish. In the wild, fish eat other small fish, shrimp, algae, insects, and insect larvae that are high in EFAs. Farmed fish are fed soy meal and other less nutritious foods which compromises the quantity of the EFAs they contain.

There has been a substantial increase in the consumption of drugs, sugar, caffeine, alcohol, and refined carbohydrates that block EFAs and their conversion to the vital prostaglandins. Combine this with the increase in toxic food, water, air, and the lack of breast feeding, and it is clear to see that the average diet does not provide enough EFAs unless a special effort is made. Not only are we are eating inadequate amounts of the right types of fats, but we are increasing our intake of the harmful ones.

Aspirin Decreases Assimilation of EFAs

Aspirin inhibits an enzyme needed to assimilate Omega-3 fatty acids into the body. Individuals using aspirin on a regular basis for arthritis discomfort or to ward off heart attack or stroke, may be doing more harm than good. EFAs are very important to help keep blood lipid levels in check and also produce anti-inflammatory prostaglandins.

How Much Fat Do We Need?

Most authorities agree that to obtain the proper level of essential fats, your total fat intake should be about 20% of your daily calories. For adults, there is no biological reason to consume saturated fats as the body can produce them if needed. So ideally, most of our fat intake would be polyunsaturated EFAs.

The daily requirement of LA (linoleic acid) is three grams per day, but our estimated optimal daily intake is between nine to 18 grams. The daily requirement of LNA (linolenic acid) is two grams per day, but an estimated optimal daily intake is at least six grams.

Many experts believe the higher the Omega-3 fatty acid consumption, the better. It may not be necessary to increase consumption of Omega-6 fats because there is usually already an over-consumption of Omega-6 from refined oils and margarine in most individuals. The average person consumes up to 20 times more Omega-6 fatty acids than Omega-3. Some authorities suggest that for optimal health, the ratio should be six-to-one, others suggest two-to-one, or even one-to-one.

Individuals with deficiency signs should double or triple the above optimal intake until the deficiency symptoms diminish. Therapeutically, individuals can safely supplement LA up to 60 grams per day and LNA up to 70 grams per day. There is no toxic dose for either fatty acid.

LA Deficiency Symptoms

Skin eruptions - Eczema
Hair loss
Liver degeneration
Behavioral disturbances
Kidney degeneration
Excessive water loss through the skin accompanied by thirst
Drying up of glands
Susceptibility to infections
Failure of wound healing
Sterility in males
Miscarriage in females
Arthritis-like conditions
Heart and circulatory problems
Growth retardation

Prolonged absence of LA from the diet can result in death. All of the deficiency symptoms can be reversed by increasing LA in the diet.

LNA Deficiency Symptoms

Growth retardation
Weakness
Impaired vision and learning ability
Motor incoordination
Neuropathy (tingling/numbing in feet, hands, arms, legs)
Behavioral changes
Elevated triglycerides
High blood pressure
Sticky platelets
Tissue inflammation
Edema
Dry skin
Mental deterioration
Low metabolic rate
Immune dysfunction

These symptoms can be reversed by increasing LNA in the diet.

EPA and DHA

Eicosapentaenoic acid (EPA) and docosahexaenoic acid (DHA) belong to the Omega-3 class of long-chain polyunsaturated fatty acids.

DHA is the longest of the long-chain polyunsaturated fatty acids - 22:6n-3. This means it has 22 carbons, six double bonds, and belongs to the Omega-3 family.

EPA is 20:5n-3, having 20 carbons, five double bonds, belonging to the Omega-3 family.

The significance of DHA as the longest of the long-chain polyunsaturated fatty acids is that it is the most sensitive to destruction and damage (mostly due to oxidation from free radicals) both inside and outside of the body. This is one of the reasons fish and fish oils have such a short "shelf life," requires refrigeration, etc,. and also why DHA can be easily destroyed in the body causing and contributing to so many various health problems.

EPA and DHA are the most commonly deficient among all the fatty acids. Under normal conditions, they can both be produced by the healthy human body if there is an adequate intake of LNA, but they are produced very slowly. In many individuals, for reasons yet unknown, this process is very inefficient. Therefore, they need to be supplied through the diet.

Supplementing DHA alone will also result an in increase in EPA levels, but supplementing EPA does not increase DHA levels.

They are both necessary to prevent fatty degenerative diseases. EPA and DHA are essential as structural components of all cell walls, are necessary for proper brain and eye development, and are required for the proper functioning of the immune, reproductive, respiratory, and circulatory systems. They are also precursors to essential regulatory prostaglandins and other eicosanoids.

Sources of DHA

DHA and EPA are usually found together in nature. They can be found in marine animals (fish, mollusks and crustaceans) and aquatic plants such as microalgae, which is where fish get their source of DHA. DHA is also found in organ meats such as liver and brain, and small amounts in lard and eggs.

For humans, fish tissues are a preferred source over fish liver which contains vitamins A and D which may be toxic in high doses. The best sources are cold-water fish such as salmon, sardines, cod, trout, and mackerel. Fish living in the wild tend to have higher levels of Omega-3 fatty acids than farmed fish. Their levels are dependent upon what foods they eat.

Additional animal sources include seals (contain about 3.5% EPA and 7.5% DHA in their fat tissue), penguins (about 3% EPA and 9% DHA in their fat tissue), polar bears (7% of each EPA and DHA), dolphins and whales (each contain 1-3% of each EPA and DHA in their fat tissue). I personally cannot imagine using these creatures for this purpose.

Other animal sources containing small amounts of EPA and DHA include zooplankton, krill, copepods, scallops, clams, oysters, and squid. These are often food for fish and marine animals providing them with their source of these EFAs.

We may also supplement EPA and DHA from microalgae (phytoplankton, algae, etc.) as an excellent vegetarian source. These tiny plants also serve as food for fish and marine animals.

Eggs contain small variable amounts of DHA. However, if eggs are overcooked or cooked improperly (breaking the yolk which exposes the sensitive fatty acids to oxygen) much of the DHA can be destroyed.

If you do take EFA supplements, it is best to divide the doses and take them with meals. Look for supplements that also have added vitamin E to reduce the risk of rancidity. Be careful with products that have vitamin A or vitamin D added because these vitamins can become toxic to the body in large doses.

SELECTED SOURCES OF DHA AND EPA

Source	LA	LNA	EPA	DHA	Total EFA	Total Fat
Fish Oils: *(Given as grams per 100 grams of oil (approx. 3 1/2 oz.) All oils are nonhydrogenated.*						
Cod (Atlantic)	1.2	0.8	12.4	21.9	41.7	100
Halibut (Pacific)	0.9	0.3	10.1	7.9	26.9	100
Mackerel	1.1	1.3	7.1	10.8	29.3	100
Rockfish	1.6	0.8	11.7	17.4	36.1	100
Salmon,						
Chinook	1.1	0.9	8.2	5.9	20.6	100
Coho	1.2	0.6	12.0	13.8	33.6	100
Sole, lemon	0.7	2.0	14.7	6.8	36.6	100
Tuna,						
Albacore	0.7	0.6	6.5	17.6	25.4	100
Bluefin	1.3	TR	6.6	20.8	28.7	100
Cod Liver Oil	1.5	0.9	8.0	14.3	29.7	100
Seafood: (*Values may vary with season, temperature, etc. Grams per 100-gram portion (approximately 3 1/2 oz.)*						
Fin Fish, fillets						
Cod (Atlantic)	N/A	N/A	0.08	0.15	0.26	0.7
Flounder	0.01	0.01	0.11	0.11	0.35	1.2
Halibut,						
Greenland	0.07	0.02	0.27	0.22	0.73	8.4
Pacific	0.02	0.03	0.11	0.20	0.55	2.0
Herring (Atl.)	0.29	0.11	0.33	0.58	1.43	6.2
Herring (Pacific)	0.12	0.03	0.76	0.57	1.67	11.1
Mackerel (Atl.)	0.14	0.10	0.65	1.10	2.44	9.8
Rockfish	0.04	0.02	0.32	0.48	0.98	3.1
Salmon,(Atl.)	0.08	0.05	0.18	0.13	0.51	5.8
Chinook	0.13	0.11	1.0	0.72	2.49	13.2

(cont.) Source	LA	LNA	EPA	DHA	Total EFA	Total Fat
Salmon,						
Coho (Pacific)	0.08	0.04	0.82	0.94	2.28	7.5
Sockeye	1.40	0.31	1.30	1.70	4.71	8.9
Trout	N/A	N/A	N/A	N/A	8	11
Tuna, canned						
Albacore	0.05	0.04	0.38	1.10	1.81	6.8
Bluefin	0.03	0.02	0.33	0.63	1.17	4.6
Shellfish						
Crab, Al. King,	0.03	0.04	0.33	0.15	0.62	1.6
Clams	0.03	0.02	0.08	0.07	0.28	1.4
Lobster	0.03	0.01	0.18	0.09	0.59	1.2
Oyster (Pacific)	0.03	0.04	0.42	0.29	0.90	2.3
Scallops	0.01	N/A	0.12	0.14	0.35	0.91
Shrimp	0.02	0.01	0.18	0.15	0.47	1.2
Snail (pond)	0.10	0.09	N/A	0.36	1.14	2.80

Americans Deficient in DHA

North Americans have some of the lowest brain DHA levels of any population on earth, due, some scientists believe, to the fact that Americans generally eat less seafood than people in other cultures. For example, studies show that the DHA level of breast milk in Japanese women is about three times that of American women.

DHA deficiencies are common among individuals with diets low in cold-water fish and in foods high in Omega-3. The decline in consumption of dietary sources such as animal organ meats and eggs has also contributed to the depressed DHA levels in the U.S. The health benefit of reducing saturated fats from these sources is generally accepted, but along with the reduction of "bad fats," "good fats" like DHA have been reduced too.

According to Udo Erasmus, author of *Fats that Heal, Fats that Kill* (Alive Books), which I highly recommend for further reading on fats, the EPA and DHA from fish take about two to three weeks to be completely metabolized in our body after being consumed. He also states that their triglyceride-lowering, artery-protecting effects last the same length of time. He therefore suggests that to maintain these protective effects, fresh fish high in Omega-3s should be eaten at least every two weeks.

Because of the many other benefits of DHA and EPA, I suspect that our intake needs may be higher than this and would recommend cold-water fish consumption once or twice a week to maintain good health.

A two-year study completed in 1989 at Royal Gwent Hospital in Wales showed that men with a history of heart disease who ate fish high in Omega-3 fatty acids twice a week, or took an equivalent amount of fish oil in the form of three capsules a day, had a 29% higher survival rate than those who didn't.

If one demonstrates signs of DHA deficiency, supplement between 1,000 to 1,500 mg. for children and 2,000 to 6,000 mg. for adults in addition to adding cold-water fish to the diet two to three times a week.

Vegetarians Have Low DHA Levels

Recent studies have shown that vegetarians are commonly deficient in DHA because they consume no animal foods and thus no preformed DHA in their diets. Blood levels of long-chain fatty acids of vegetarians show that DHA levels are very low, especially in long-term vegetarians. (Agren, 1995)

A vegetarian diet in mothers can also effect the fatty acid levels of newborns. Researchers found the DHA levels of breast-fed infants of vegetarian mothers to be only about one-third the level of breast-fed infants of non-vegetarian mothers. This suggests that the vegetarian mother's DHA levels might be low. (Sanders)

This may be an indication that strict vegetarians may need to supplement their diets with DHA. This is especially important for vegetarian women of childbearing age, pregnant mothers, and nursing mothers.

Plant-based DHA supplements are now available to help meet the fatty acids needs of individuals who choose to avoid fish and also for those who are allergic to fish.

Who Should Supplement DHA?

■ **Individuals who limit their meat and egg intake** (such as vegetarians or those on low-cholesterol programs) and others whose diets are generally low in DHA can assure an adequate supply by supplementing their diets with DHA.

■ **Individuals showing signs of DHA deficiency:**

Insufficient brain and vision development in infants

Visual impairment/Visual blurring

Abnormal electroretinogram (ERG)

Impaired learning/Memory

Numbness in fingers, hands, toes and feet

Inflammatory skin disorders: Eczema, Dermatitis, Psoriasis, etc.

Behavior disorders: ADD/ADHD, Addictions, Alcoholism, Violence, Aggression, etc.

Neurological disorders: Depression, Alzheimer's disease, Memory loss, Dementia, Schizophrenia, Dyslexia, etc.

Reduced DHA in breast milk

■ **Pregnant or lactating mothers**

Adequate dietary intake of DHA is particularly important for pregnant and nursing mothers. Significant brain and eye development occurs in utero and continues during the first year after birth. Infants rely on their mothers to supply DHA for the developing brain and eyes initially through the placenta and then through breast milk. DHA is the most abundant Omega-3 long-chain fatty acid in breast milk and studies show that breast-fed babies have IQ advantages over babies fed formula without DHA. But, DHA levels in the breast milk of U.S. women are among the lowest in the world.

■ Individuals who are allergic to fish

Individuals who suffer from allergic reactions from fish can assure a safe and adequate supply by supplementing their diets with algae-sourced DHA.

■ Individuals who cannot convert LNA into DHA

Some individuals are deficient in (or have malfunctioning of) the enzyme that converts dietary LNA into DHA, called delta-6-desaturase. This enzyme also converts LNA to PGE3. LA and LNA compete for this enzyme so if there is an excess of one over the other, that one "wins" and products of that fatty acid predominate.

Factors which inhibit delta-6-desaturase enzyme activity include:

* High intake of saturated fats and trans-fatty acids (which act like saturated fats in the body)

* Stress

* Diabetes (elevated insulin)

* Excess sugar consumption (which increases insulin and also converts to saturated fats)

* Obesity

* Anti-inflammatory agents including aspirin, ibuprofen and corticosteriods

* Infancy (before the age of one, this enzyme is inac-

tive which is why preformed DHA, AA, and GLA are important parts of an infants diet.)

* Aging (After the age of 20 the efficiency of the enzyme declines.)

* Certain genetic factors causing allergies and eczema (called atopy) retinitis pigmentosa, and other genetic errors

■ Individuals who consume alcohol

It is true. Alcohol kills brain cells. Alcohol crosses the blood brain barrier, causing free radical damage to the brain cell membranes which are largely made of sensitive DHA. If inadequate dietary DHA is present to help replace the damaged cells, alcohol can literally destroy the brain leading to memory loss, and eventually dementia in long-term alcoholics. Alcoholics who have progressed to liver disease have even lower levels of DHA.

Helpful suggestions to help protect DHA and other long-chain polyunsaturated fatty acids in the body:

■ Avoid free radicals which are caused by stress, cigarette smoke, car exhaust, ozone, radiation, alcohol, certain drugs, and other toxins.

■ Avoid saturated fats and excess AA both found in animal products such as meat, dairy and eggs.

■ Supplement antioxidants such as vitamin E, vitamin C, ginkgo biloba, alpha lipoic acid, CoQ10, N-acetyl cysteine, grape seed extract, pine bark extract, bilberry, and other anthocyanadins, etc.

DHA: The Brain, Nervous Tissue and Vision

DHA is most concentrated in the brain and nervous system. Its highest concentration is in the parts of the brain that require a high degree of electrical activity. DHA is a very long, electrically active fatty acid. Its six double bonds give it a curved, spacious shape that provides for a fluid, supple nerve cell membrane. Other fatty acids do not appear to share these unique properties of DHA. The spatial aspects of DHA hold the receptors in place within the nerve cell membranes. Proper binding of neurotransmitters with receptors is essential for optimal nerve communication.

When the brain does not receive enough DHA, it uses docosapentaenoic acid (DPA) as a replacement fatty acid. This is an Omega-6 fatty acid which is considered largely inadequate. (Stubbs)

Where High Concentrations of DHA Are Found:

■ **Cerebral cortex:** This is the outer layer of the brain which is dense with cells rich in DHA.

The brain is very special and unique. It is composed of 60% lipids, and it has a very high rate of energy con-

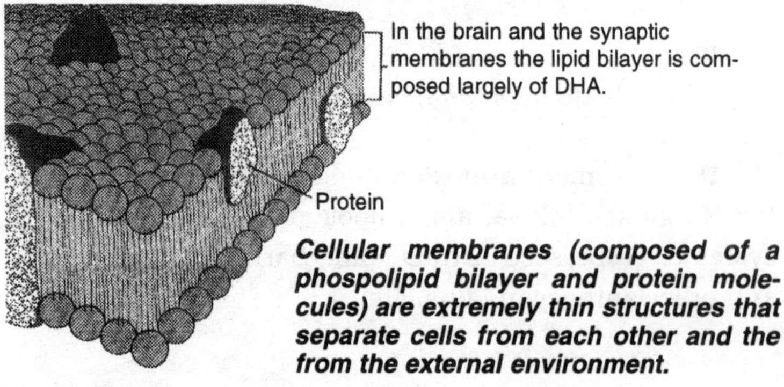

Cellular membranes (composed of a phospolipid bilayer and protein molecules) are extremely thin structures that separate cells from each other and the from the external environment.

sumption. Energy consumption produces damaging free radicals which can injure the sensitive cellular membranes.

■ **Synaptic membranes:** The endings of nerve cells where messages are transferred. *(see figure below.)*

Nerve cells (neurons) transfer information between themselves through chemicals called neurotransmitters which migrate between the space between neurons called the synaptic clept.

■ **Mitochondria of nerve cells:** These are the tiny bodies that generate all of the energy for all cells and that gives the brain life. All organs of the body are exposed to DHA as it circulates through the bloodstream, but the requirement of the brain is so high, and its function there so important, that the brain takes up most of it.

■ **Photoreceptors:** This is the portion of the retina of the eye that is lined with a delicate, highly specialized fatty membrane that uses DHA for conduction. This area receives light stimulation and is vital for proper vision. The retina of the eye has the highest concentration of

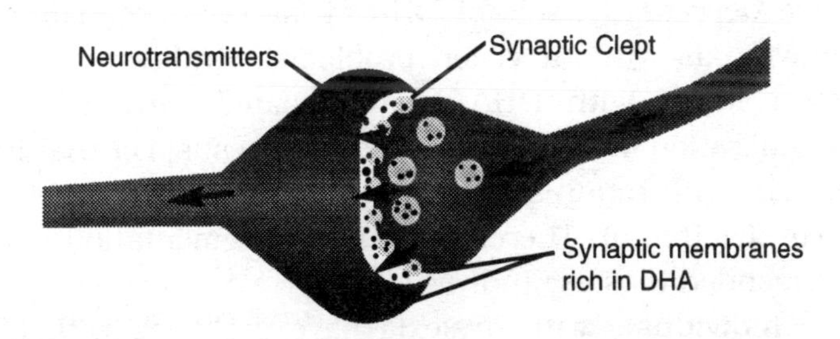

Nerve implulses (transmitter substances) travel through synaptic membranes, where the concentration of DHA is among the highest in the body. Without adequate DHA, impulses are not transferred with optimal efficiency.

There may be over 20,000 synapses along a single nerve in the body. The number of synapses is more important in terms of intelligence than the number of nerve cells in the brain.

DHA of any tissue in the body. The retina is a frail nervous tissue membrane of the eye. Receptors in the retina receive light and quickly translate that information into diverse signals. These impulses are then sent to regions of the brain to interpret them.

The fatty acid DHA is crucial for formation of the retina in the fetus and infant. Infants who received adequate DHA have better visual acuity and image processing than those not receiving adequate DHA. It is therefore necessary for proper visual function.

A constant supply of DHA is needed throughout life in order to preserve vision. The retina's photoreceptor cells are shed and replaced with new cells daily. In order to retain vital visual capacity, DHA must be provided to the newly forming cells.

DHA is also believed to improve visual function in the elderly. DHA levels in the retina decrease with age. After the age of 20, the delta-6-desaturase enzymes needed for us to produce DHA from LNA begin to wear out. Therefore, a regular supply of DHA through the diet or supplementation becomes increasingly important.

Individuals with visual disorders such as retinitis pigmentosa (a degenerative disease of the retina) are found to have depressed levels of DHA (40% lower) compared to individuals without vision problems. Studies show that lower levels with DHA are associated with increased degeneration of the retina. Researchers suspect that individuals with retinitis pigmentosa cannot convert LNA into DHA. (Hoffman) Therefore, DHA supplementation is recommended in these individuals.

Individuals with dyslexia have problems with their retinas, and also have difficulty processing visual information in the brain. Some researchers, such as B. Jacqueline Stordy of the University of Surrey, Guildford, UK, have reported improvement in reading ability and behavior in dyslexic individuals with DHA supplementation, but additional trials are needed to determine to what extent.

Prostaglandins

Much of the effects of the essential fatty acids are a result of their conversion into powerful hormone-like substances in the body known as prostaglandins (PGs). Prostaglandins are thought to play a role in the regulation and function of every organ and cell in the human body. Their wide-ranging effects help explain the multitude of diverse properties of the essential fatty acids.

Prostaglandins are made in the body from essential fatty acids. They are short-lived chemicals that regulate cellular activities on a moment-to-moment basis. Chemically, PGs are made through very specific enzyme-controlled oxidation of highly unsaturated fatty acids.

Over 30 different PGs have been isolated and identified. Each has different and highly specific functions, and some PGs are stronger in their function than others. Prostaglandins fall into 3 families or series, according to which fatty acid they were made from.

Series 1 and 2 prostaglandins come from the Omega-6 oils. Linoleic acid (LA) is converted in the body into gamma-linolenic acid (GLA), then to dihomo-gamma-linolenic acid (DGLA), and then to arachidonic acid (AA). An excess intake of LA can convert into AA. AA also comes from animal products such as meat and dairy,

Series 1 prostaglandins are made from DGLA.

Series 2 prostaglandins are made from AA.

Series 3 prostaglandins are made from the Omega-3 oils. Alpha-linolenic acid (LNA) is converted by the body to stearidonic acid (SDA), then to eicosatetraenoic acid (ETA), then to eicosapentaenoic acid (EPA). Series 3 prostaglandins are made from EPA.

Functions of Prostaglandins

Actions of PGE1 - One of the "good guys"

* In our kidneys, it helps remove sodium and excess fluid from our body as a diuretic.

* Relaxes blood vessels, improving circulation, lowering blood pressure, and relieving angina.

* Improves nerve function, increases feeling of well-being.

* Regulates calcium metabolism.

* Improves the functioning of T-cells in our immune system, which destroy foreign molecules and cells.

* Prevents the release of AA from our cell membranes.

* Required for proper functioning of the immune system- activates T-cells, which destroy cancer and other unwanted substances in the cells of the body.

* Inhibits cell proliferation and normalizes malignant or mutated cells, thereby promotes cancer cell reversal.

* Decreases inflammation response. It is effective against inflammatory conditions such as eczema and arthritis as well as auto-immune diseases including rheumatoid arthritis. Drugs commonly prescribed for these diseases, however, deactivate PGE1. In addition to its own anti-inflammatory action, GLA/PGE1 controls the release of stored AA, thereby further reducing potential pain and inflammation.

* Protects against various forms of heart and vascular disease including stroke, heart attack, and arterial deterioration. It may slow down cholesterol production and is a vasodilator which controls blood pressure, and inhibits blood clotting, a major cause of thrombosis, stroke, and cardiovascular disease.

* Regulates brain function and nerve impulses; clinical trials show GLA to be helpful in the treatment of schizophrenia.

* Often alleviates the "dry eye" (Sjogren's and sicca) syndrome - the inability to produce tears.

* Is a "human growth factor" which stimulates growth that has been retarded.

* Regulates the action of insulin and is therefore beneficial in diabetes; appears to minimize damage to the heart, eyes, nerves, and kidneys in all forms of diabetes.

* Multiple sclerosis (MS) is thought to result in part from LA not being converted to PGE1.

* Speeds up the metabolism in those with stagnancy and obesity. Is often used to assist weight loss.

PGE1 Enemies

Alcohol consumption temporarily raises PGE1 levels dramatically, but depression occurs afterward (hangover), which is cured by GLA/PGE1. Alcoholics are deficient in PGE1, as alcohol creates failure of the body to produce it. GLA is effective in helping to reduce alcohol cravings and also helps restore liver and brain function in alcoholics.

Certain nutrient deficiencies, intoxicants, synthetic drugs (including aspirin), excess saturated fat, and other conditions (age, genetics, etc.) limit the production of PGE1.

Prostate problems, premenstrual syndrome (PMS), cystic mastitis (breast lumps), brittle nails, and hyperactivity in children often result from low levels of PGE1.

Actions of PGE2 - the "bad guy"

PGE2 opposes PGE1 functions. In most cases of everyday living, it has negative effects in the body. Its value lies under primitive "fight or flight" conditions, not normally experienced in modern-day society.

One member, PGl2, acts like PGE1, helping to keep platelets from sticking together. But another member, called PGE2, promotes platelet aggregation, the first step in clot formation. PGE2 also induces the kidney to retain salt, leading to water retention and high blood pressure. It also causes inflammation.

Arachidonic acid (AA) causes prostaglandin production of the type PGE2. In excess, they can produce pain and inflammation, and encourage blood to clot. In a sense, AA works against the beneficial effects of both GLA and LNA. Excess AA in the human body is largely from the consumption of animal products, or can also be the result of an excess intake of LA.

Aspirin and various steroid drugs block the production of PGE2 and therefore, reduce clotting, pain, and fever. Many doctors advise their patients to take aspirin to protect the heart and for the pain of arthritis, but aspirin also blocks the production of the beneficial PGE1, so that when it is used for arthritis and heart disease, the inflammation and deterioration of tissue from leukotrienes continue. Instead, one may consider increasing the production of PGE1 and PGE3. Remember, PGE2 production is limited by increases in either PGE1 or PGE3; and both PGE1 and PGE3 have their own anti-inflammatory action and many other valuable properties.

Having adequate levels of PGEl helps keep AA stored in cell membranes where it cannot be converted into series 2 prostaglandins. This prevents the bad effects of PGE2 from occurring. The good effect of PGI2 in preventing platelet stickiness is already covered by PGE1 and also by series 3 prostaglandins.

Actions of PGE3 - One of the "good guys"

Two members of this series, called PGE3 and PGl3, have very weak platelet stickiness (aggregating) effects. The most powerful effect of PG3s is not so much in their specific action, but in the fact that EPA, their parent, prevents AA from being released from membranes, thereby preventing "bad" PG2s from being made. EPA is the single most important factor limiting PGE2 production and explains why fish oils can prevent degenerative cardiovascular changes, water retention, and inflammation caused by excessive PG2s.

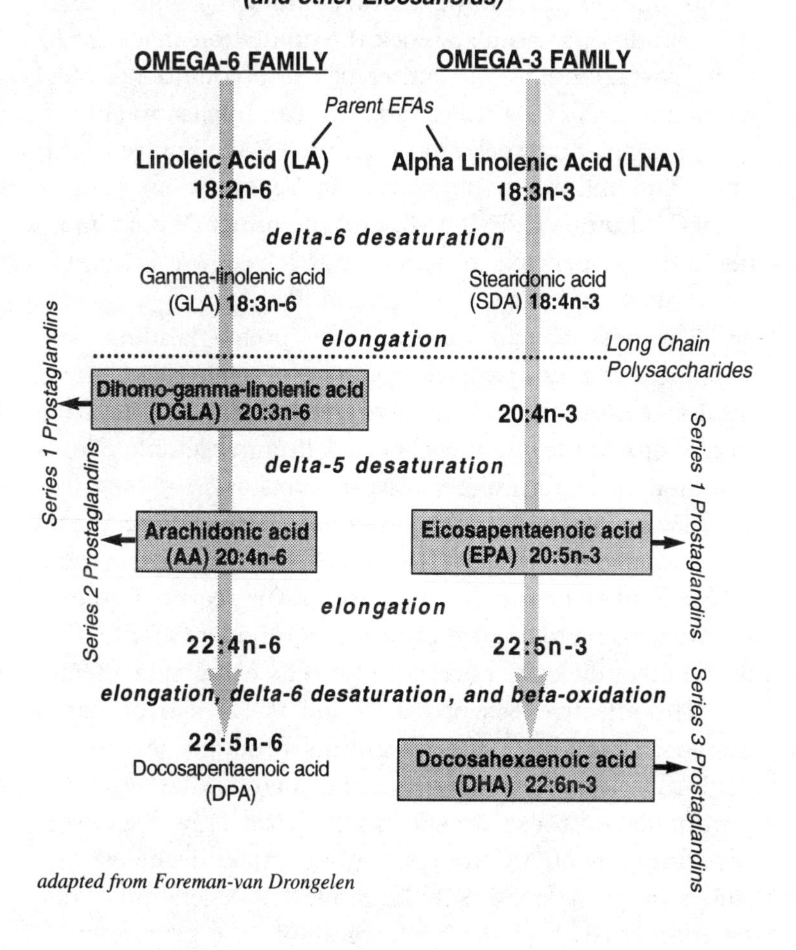

adapted from Foreman-van Drongelen

Prostaglandin Regulation

Healthy individuals make the prostaglandins they need from LA and LNA, if there are adequate levels in the body. It cannot if there is an inadequate consumption of these EFAs, and imbalances can occur if there is an excess intake of LA compared to LNA, which is very common.

In addition to improper nutritional conditions, metabolic conditions can also block the ability to convert EFAs into prostaglandins. In either of these conditions, supplementation of GLA, EPA and DHA can bypass the block.

Prostaglandin production from EFAs requires vitamins and minerals that can also be taken as supplements. Vitamins C, B-3, B-6, and the minerals zinc, magnesium and perhaps others, are involved. We still don't know all we need to in this area of health.

The best way to supply "good" prostaglandins is to support the body's natural production through a balanced intake of EFAs, especially EPA and DHA and minimizing intake of AA. Elevated levels of LA (from vegetable oils) is common and can further elevate levels of AA.

By giving the body the materials it needs to make its own supply of healthy prostaglandins, in the required location, at the required time, and in the required quantities, according to its own internal requirements for health, is the most effective means and produces no side effects.

The effectiveness of PGE1 and PGE3 can be easily increased by decreasing all animal products (except for cold-water fish) and avoiding the factors that limit the production of these prostaglandins. Not only does overconsumption of AA from saturated fat and cholesterol-rich animal products set the stage for degenerative disease processes, but it directly inhibits the generation of the invaluable prostaglandins from GLA and Omega-3 oils.

Pharmaceutical preparations of prostaglandin-like

substances are not "perfect" enough to act correctly in the body and without serious side effects. Prostaglandins cannot be directly consumed orally as they are destroyed during digestion. Large injected doses of prostaglandins are required as the body breaks them down before reaching target tissues.

Substances Produced From Arachidonic Acid (AA)

Elevated levels of AA in the cell membranes can lead to the production of the following detrimental substances. DHA supplementation has shown to reduce the production of all of these. It would also be helpful to reduce the intake of AA which is found predominantly in animal fat.

■ **Prostaglandin E2,** which causes inflammation and pain.

■ **Leukotrienes,** which can act beneficially to heal wounds and injuries, but in excess are thought to provoke breast lumps and inflammations such as arthritis (see segment on arthritis). The inflammation of rheumatoid arthritis may be a direct result of excessive leukotrienes. Other leukotriene-related conditions are asthma, dermatitis, allergic rhinitis (hay fever), psoriasis, and lupus (systemic lupus erythematosus). AA-derived PGE2 also stimulates cell division and proliferation, which, when taken to an extreme degree, can be directly linked to cancer and tumors.

■ **Tromboxane A2,** which causes clotting (increases platelet stickiness) and blood vessel spasms.

■ **Prostacyclin,** which reduces platelet stickiness and relaxes blood vessels.

PROSTAGLANDIN FORMATION FROM FATTY-ACID SOURCES

PGE1

Linoleic Acid (LA) from nuts, seeds, grains, legumes, most vegetables, fruit, and animal products.

Gamma-Linolenic Acid (GLA): Direct sources: mother's milk, spirulina, oils from evening primrose, black currant, and borage seeds, seeds of maple, sycamore, and related trees.

PGE2

Arachidonic Acid (AA) from animal meats, dairy, and eggs.

PGE3

Alpha-Linolenic Acid (ALA) found in the seeds of flax, chia, pumpkin, also walnuts, soy products, and dark greens. Cold-climate crops contain relatively more ALA.

EPA and DHA: Direct sources: fish such as salmon, sardine, tuna, lake trout, eel, anchovy, pilchard, and butterfish, microalgae, mother's milk, and most snakes.

Infant Nutrition Needs

DHA is critical for the developing fetal brain and during infancy. It is essential for brain and eye development and for mental and visual function. Children who receive adequate DHA have been shown to have better visual acuity and IQ than children who receive inadequate amounts of this fatty acid. Preformed DHA must be provided since the body during these periods in unable to make it even if there is adequate LNA. Independent studies have shown that unless preformed DHA is provided to infants, their brain DHA levels are subnormal compared to breast-fed babies. Studies have clearly shown that DHA supplemented pre-term infants have higher red blood cell DHA and better visual acuity than standard formula (without DHA) fed pre-term infants. (Cockburn, Carlson)

Experimental evidence demonstrates that the effect of EFA deficiency during early development is damaging and permanent. The risk of neurodevelopmental disorders is highest in the very-low-birth-weight babies who are most likely to have been born to undernourished mothers. These babies tend to be born with AA and DHA deficits.

Humans first obtain DHA through the placenta and from breast milk. Mother's milk contains vastly more DHA than cow's milk, making mother's milk a far superior source of this nutrient for the infant and toddler brain.

DHA - Most Abundant Structural Fat in the Brain

DHA and AA are of critical importance as the basic building blocks of each cell formed. Brain tissue is about 60% structural fat, of which about 25% is DHA. Brain and other nervous tissues are unique in containing this high concentration of DHA.

The rate of brain growth in the perinatal period is so

rapid that the baby's capacity to synthesize DHA from LNA is insufficient to keep up with the demand by the growing brain and nervous system.

It has been known for some time that deficiencies of Omega-3 fatty acids leads to deficits in learning abilities. (Crawford)

Brain Growth and the Mother

Who doesn't want their child to be brilliant? Who doesn't want their child to grow up to get good grades, graduate at the top of their class, be a Nobel Prize Winner or work in some prestigious profession? As a mother, you may be much more responsible for this determination than you realize, and the effects may be complete before you even realize that you are pregnant.

Some 70% of the total number of brain cells to last an individual's lifetime have divided before birth. In fact, the most active period of brain cell division is in the very first few weeks of embryonic development. If you are "trying" to get pregnant, for the health of the baby, for several months before the fact you should be taking dietary precautions as if you were already pregnant. The neurological development of the baby depends upon it.

Studies show that DHA supplemented infants performed significantly better than controls on the Bayley mental scale where perception, memory learning, problem solving, vocalization, early verbal communication and abstract thinking are tested. (Cockburn, Carlson)

DHA Means Improved Mental Function in Infants

Studies show that children who were breast-fed perform better on cognitive function tests later in life (by five to nine IQ points) than those who were formula-fed even after taking into account all confounding factors associated with developmental test performance (e.g., socio-economic status, IQ of parents, etc.). A greater developmental disparity has been established for low-weight pre-term infants born without the benefit of the maternal delivery of DHA during the last trimester. They experience deficits of up to 20 IQ points compared to term infants and are at greater risk for behavioral problems. Evidence points to DHA as the reason.

DHA - Essential for Visual Development

DHA makes up about 60% of the rod outer segments of the retina of the eye. The retinal photoreceptor cells in our eyes contains the highest concentration of DHA in the entire body. Because of this DHA-containing lipid layer, membranes are allowed to move freely across one another which is necessary for the retinal cell to quickly switch "off and on." DHA, and also EPA, are involved with the conversion of light energy entering our eyes into the chemical energy of nerve impulses.

The development of the visual system of human infants is clearly dependent in part on the availability of DHA. The effect is most apparent in pre-term infants. Research shows an increase in the prevalence of retinopathy in premature infants. (Crawford) This suggests that an exogenous source of DHA is crucially needed during the time of rapid development of the retina. (Heird)

EFA Needs During Pregnancy

An increased EFA dietary requirement is present throughout the nine months of pregnancy to provide for the fat storage in the early trimester and the growth in the later trimesters. Based on accumulated data, researchers have estimated that approximately 2.2 grams per day of EFAs is needed to meet these needs in a normal pregnancy.

Studies suggest that a relative deficiency of long-chain polyunsaturated Omega-3 fatty acids develops during pregnancy. (Holman) The relationship between AA and birth weight and DHA and gestational age are consistent as an indicator of maturation with DHA intakes. The evidence suggests that a diet high in fish is associated with longer gestation, higher birth weights and reduced incidence of premature births. (Leaf)

EFA Needs During Lactation

The diet of a well-nourished mother provides adequate EFA to the infant. Fat output is the most variable component of milk and this depends on the maternal nutrition and secretion of the hormone prolactin.

During the first three months of lactation, the mother's diet should contain an additional three to four grams per day of EFA. As fat supplies deplete, this should increase up to five grams per day.

Researchers investigating the content of Omega-3 fatty acids in mother's milk found levels to vary to a large extent - likely due to the large variations in diets. There are not enough studies to determine what are the optimal levels. Obviously women on a high marine diet have significantly higher DHA levels compared to women on a vegan diet. Women in the U.S. tend to have lower levels of DHA in breast milk compared to other countries, about one-third the level found in Japanese women.

Percentage of Fats in Human Breast Milk

Country	LA	LNA	EPA	DHA	Total EFA	Total Fat
Australia	0.4	0.02	0.01	0.01	0.5	4
North America	0.6	0.04	0	0.01	0.73	4
Malaysia	0.4	0.01	N/A	0.04	0.46	4

DHA is a very important component of breast milk. Women should breast-feed as long as possible to ensure that their babies get an adequate supply. Women also should review their diets to ensure that their breast milk contains enough of this nutrient. Dietary trends indicate that DHA intake in the U.S. has declined by about 50% over the last 50 years.

Breast-fed infants have significantly higher concentrations of DHA in the cerebral cortex phospholipids than infants fed infant formula without DHA. There is a large body of evidence showing that breast-fed infants test higher in cognitive development scores and demonstrate higher IQ levels. Breast-fed infants are also known to be more active and to pass their motor milestones earlier than formula-fed babies without DHA. (Taylor, Fergusson)

Also interesting is that children who never received breast milk were perceived as having worse behavior (at five years old) than breast-fed infants. However, there seemed to be no difference is speech development or speech problems in the two groups. (Taylor)

This research conducted over 10 years ago still holds true for more recent studies. By the early 1990's, many formula developers, showing increased interest in EFAs, were supplementing formulas with LA and LNA. Unfortunately, it was later revealed that infants were unable to effectively convert LNA into DHA, and thus were still deficient compared to breast-fed babies. (Neuringer)

WHO Committee Recommends DHA in Formula

To ensure proper brain development, some scientists advocate supplementing infant formula with DHA. An expert committee of the World Health Organization (WHO) has recommended that all pre-term and term infant formulas contain DHA at levels found in human breast milk. The British Nutrition Foundation and the European Society of Pediatric Gastroenterology and Nutrition have made similar recommendations. DHA is already in some European and Asian formulas, but at this writing, not yet in the United States.

Several major infant formula manufacturers, including Wyeth-Ayerst (American Home Products), Mead

Johnson Nutritional Group (Bristol-Myers Squibb), Nutricia NV, Sandoz, and Maabarot Products have licensed the technology to incorporate DHA into their products. These infant formulas are now available in Europe.

Research data demonstrates that DHA and AA are absorbed by infants from supplemented formula at least as effectively as from human milk. The net absorption depends on the amount of dietary intake, and seems to be influenced by the dietary LCP source. (Boehm, Carnielli)

In three randomized, double-blind clinical trials, pre-term infants were fed typical pre-term or term formulas and experimental formulas supplemented with Omega-3 and Omega-6 long-chain fatty acids.

Pre-term infants fed both AA (0.43%) and DHA (0.1%) had better AA status than controls. This is significant because of the evidence that AA status and the Omega-6/Omega-3 rates are related to growth of pre-term infants. (Carlson)

Down Syndrome

Down syndrome is a genetic condition characterized by an extra chromosome 21. People with Down syndrome often suffer from learning problems and commonly go on to develop a dementia that is similar to Alzheimer's disease. The extra chromosome causes an elevated level of an enzyme called SOD that contributes to a high level of rancid fatty acids, called lipid peroxides, in the brain. Because of this free radical stress, antioxidant nutrients are depleted and fatty acids and brain tissue are destroyed. This suggests that in Down syndrome, balancing essential fatty acids and antioxidants can be critical to preserving brain function.

The recommended supplemental DHA dosage for non-nursing infants, is 20 mg. per kilogram (2.2 lbs.) body weight per day.

Cerebral Palsy

Cerebral palsy describes a number of different brain disorders that occur in the first few developmental years of life. It includes a wide range of symptoms. It often occurs from events at birth, but may be closely linked to what happens during pregnancy.

The rate of cerebral palsy is highest in premature babies and has increased significantly among low-birth-weight babies. Premature or low-birth-weight babies do not have the continued supply of brain fatty acids from the placenta. According to Dr. Michael Crawford, the DHA levels of a baby born prematurely fall quickly to only one-fifth of the level of the placenta. The level of DHA fell more in those infants with intrauterine growth retardation, Unless the child immediately begins to receive DHA there may not be adequate brain-fats to properly form the developing nervous system. (Leaf)

Some doctors suggest cerebral palsy may be, in part, due to deprivation of key brain fatty acids such as DHA. (Leaf, Crawford) Supplementation with fatty acids may improve some functions and lead to an improved quality of life.

Attention-Deficit Hyperactivity Disorder

Attention-deficit hyperactivity disorder (ADHD) is the term used to describe children who are inattentive, impulsive, and hyperactive. The cause is said to be unknown and is thought to be due to multiple factors. Some researchers hypothesize that some children with ADHD have altered fatty acid metabolism.

EFA Levels Low in Hyperactive Children

Researchers investigating essential fatty acid metabolism in boys with attention-deficit hyperactivity disorder at Purdue University, found that 53 subjects with ADHD had significantly lower concentrations of DHA and AA than did the 43 control subjects. Also, a subgroup of 21 subjects with ADHD exhibiting many symptoms of essential fatty acid (EFA) deficiency had significantly lower plasma concentrations of DHA and AA than did 32 subjects with ADHD with few EFA deficiency symptoms. The precise reason for lower fatty acid concentrations in some children with ADHD is not clear. (Stevens)

Another study comparing 48 hyperactive children with 49 age-and-sex-matched controls also found depressed levels of EFAs, specifically DHA, dihomogammalinolenic (DGLA), and AA. They also found that significantly more hyperactive children had auditory, visual, language, reading, and learning difficulties, and low-birth-weight than the controls. Serum results of EFA levels compared to controls were as follows: (Mitchell)

EFA	Hyperactive	Control
DHA	41.6 mg/ml serum	49.5 mg/ml serum
DGLA	34.9 mg/ml serum	41.3 mg/ml serum
AA	127.1 mg/ml serum	147.0 mg/ml serum

Abnormal Conversion of EFAs Suspected

The Hyperactive Children's Support Group in England has researched and also found the connection between ADD and the deficiency of EFAs. Their research led them to suspect that hyperactive children might have a problem with an important pathway in the body that converts EFAs to prostaglandins-hormone-like substances that control all bodily functions at the cellular level.

Their research demonstrated that many children cannot metabolize nor absorb EFAs normally. The EFA requirements of these children are thus higher than normal. An EFA deficiency results in some serious symptoms, many of which look like ADD. Other common symptoms are eczema, asthma, and allergies.

The following points suggest the involvement of EFAs in ADD-like behavior:

■ Hyperactive male children outnumber females by three to one. They required two to three times more EFAs than females to prevent the signs of essential fatty acid deficiency. Males may have more difficulty converting EFAs to prostaglandins.

■ About two-thirds of children with EFA deficiencies have abnormal thirst. Thirst seems to be a key feature.

■ Asthma, eczema and other skin problems are more common in hyperactive children due to the defective formation of prostaglandins.

■ Many children with ADD are zinc deficient. Zinc is required for the conversion of EFAs to prostaglandins.

■ Wheat and milk can adversely affect some children. These foods can block the conversion of EFAs.

Fatty acids must be converted in order to be used. Diet plays a key role in the process because it takes adequate zinc, vitamin E, vitamin C, niacin, and pyridoxine (B-6) to convert Omega-6 fatty acids to Omega-3 fatty acids. Other factors that can interfere with this important conversion process are chronic illness, stress, and eating large quantities of saturated fats and hydrogenated oils which contain trans-fatty acids.

Hydrogenated fats are found in margarine and shortening. These are found in most processed foods such as bread, rolls, crackers, pies, pretzels, cookies, donuts, bread sticks, muffins, bread crumbs, stuffing, pop tarts, biscuits, pancake mix, quick breads, potato chips, candy bars, non-dairy creamers, peanut butter, salad oils, fast-food shakes, baked and canned goods, packed-in-oil products, and fried foods at restaurants. The annual consumption of trans-fatty acids is almost twice the total intake of all other unnatural food additives put together.

Experts list a large number of natural food substances and unnatural food additives that may precipitate symptoms in children with ADD. Many of these are known inhibitors (or are chemically related to known inhibitors) of the conversion of EFAs to prostaglandins.

These include natural salicylates and other coloring materials such as yellow dye (also found in margarine and many processed foods). Yellow dye is known to inhibit prostaglandin formation when EFA levels are very low, but has little or no effect when EFA levels are high. This suggests the possibility that children with normal levels of EFAs are not affected. Adding EFAs to the diet of children with hyperactivity may correct the symptoms.

Elimination Of Fat Is Not The Answer

Parents are sometimes told that all fats and oils are bad and often give the children inadequate amounts. Non and even low-fat diets may be very dangerous for infants and children. This can also actually aggravate hyperactivity symptoms.

Fat provides an important source of concentrated calories needed for growth. Calcium and fat soluble vitamins (A, E, and D) require fat for absorption. Growing children require more fat and oil in their diet than adults.

Do Children Outgrow ADD?

Without finding and treating the underlying cause of the symptoms, most individuals do not outgrow ADD. By the time they become adolescents, they often appear to be less hyperactive, but may still have many of the symptoms of ADD. Because of their poor learning skills and under-developed social skills, they continue to do poorly in school.

Half of all children with ADD still show signs of the problem into adulthood. Approximately 80% of children with ADD will still meet the criteria for this diagnosis when adolescents. Most of them usually go on to become adults with ADD.

Addiction and Other Behavior Disorders

The Cortisol Connection?

The relationship between DHA, ADD, addiction and other behavior disorders is of particular interest. Researchers have shown us another intriguing common thread between these groups of individuals: elevated cortisol levels. Cortisol is a steroid hormone, produced by the adrenal glands, often due to stress. Elevated levels are associated with sleep difficulties (one cannot enter REM stage), and catabolism - the breakdown of complex substances (like muscle protein). Catabolism is the opposite of anabolism, which is often associated with muscle building.

While I could not specifically find studies on individuals with alcoholism, ADD or other similar disorders on the effects of EFAs on cortisol levels, I did find several studies showing that supplementing EFAs increases the resistance of cells to cortisol.

Researchers at the University of Toronto demonstrated that cortisol-sensitive lymphocytes have a much lower capacity than cortisol-resistant cells to break down cortisol and that LA inhibits the break-down of cortisol by lymphocytes and changes the sensitivity of lymphocytes to cortisol. (Klein)

Studies conducted to study the effects of DHA and EPA on patients with elevated blood triglyceride levels showed that levels of serum phosphatidylcholine were also increased. The drug-treated group received four grams per day of an 85% concentrate of EPA and DHA for six weeks. In the treated patients, mean increases of 160% and 300% in phosphatidylcholines were observed compared to no increase in the placebo group. Also, in the treated group, triglycerides were reduced 26%, while they increased by 7% in the placebo group. (McKeone)

Phosphatidylcholine levels are significant because this important phospholipid is believed to reduce the negative effects of elevated cortisol. (Monteleone)

Low DHA Levels Seen in Alcoholics, and those with Violent and Impulsive Tendencies

Researchers have uncovered a common trend among essential fatty acids and prostaglandins in alcoholic, habitually violent, and impulsive offenders... low DHA levels.

Researchers at Helsinki University Central Hospital in Finland measured essential fatty acid levels among habitually violent and impulsive male offenders, who all had alcohol abuse problems, compared to nonviolent control persons. They found the following:

- Low levels of LA in intermittent explosive disorder.
- Elevated levels of dihomogammalinolenic acid (DGLA) and some subsequent Omega-6 acids were elevated among all offenders.
- Elevated levels of Oleic acid.
- High DGLA which correlated with low cholesterol level in individuals with intermittent explosive disorder.
- The AA metabolites PGE2 and TxB2 were elevated in individuals with violent antisocial personality.
- The PGE1/DGLA ratio was low in individuals with intermittent explosive disorder.

■ The number of registered violent crimes and violent suicidal attempts correlated with high phospholipid DGLA values.

The researchers suggested the possibility that the high phospholipid DGLA is connected with low free DGLA pool, and contributed to low PGE1 formation. (Virkkunen)

DHA Found to Reduce Aggression

Japanese researchers examined the effects of DHA on aggression in a double-blind placebo-controlled study on students. Forty-one students took either DHA capsules containing 1.5-1.8 grams DHA/day or control soybean oil capsules for three months. They took a psychological test (P-F Study) and Stroop and dementia-detecting tests at the start of the study at the end of summer vacation and at the end of the study in the middle of high mental stress during final exams.

In the control group, aggression against others according to results in the P-F Study was significantly increased at the end of the study as compared with that measured at the start, whereas it was not significantly changed in the group supplementing DHA. The differences between the DHA and control groups were 16.8 to -3.0%. DHA supplementation also did not affect the Stroop and dementia-detecting tests. Thus, DHA intake effectively prevented aggression towards others from increasing at times of mental stress. The researchers suggested that this finding might also help us understand how fish oils prevent disease like coronary heart disease. (Hamazaki)

Alcoholism

Alcoholics represent 5-10% of the U.S. population. Its causes are varied. Alcoholism (dependence on alcohol, marked by bad behavior) is a long-term illness that starts slowly and may occur at any age. The most frequent medical dilemmas include mental problems and breakdown of the liver (cirrhosis).

Chronic Alcoholism Decreases AA and DHA

Researchers examined fatty acid levels in two groups of alcoholic subjects, one of them with chronic liver disease, and compared them to a control group of healthy subjects. They found the following levels in both groups of alcoholic patients:

DHA	significantly reduced
Linoleic acid (LA)	unaffected
Arachidonic acid (AA)	reduced
Docosatetraenoic acid (DA)	reduced

The decreases were more significant in the presence of chronic liver disease. The researchers concluded that chronic alcohol ingestion causes important changes in long-chain polyunsaturated fatty acids, and that these changes are exacerbated when patients suffer from chronic liver disease. (Pita)

Another group of researchers had also found levels of AA, DA and DHA to be significantly reduced compared to non-alcoholic controls. A high level of bilirubin was also seen in most patients. Bilirubin is the orange-yellow pigment formed by the breakdown of hemoglobin in red blood cells. It is normally excreted into the bile and eliminated from the body. A buildup causes the characteristic yellow skin of jaundice. The researchers discussed the possibility of the increased levels due to membrane changes in the red blood cells as a result of low DHA and AA. (Adachi)

When cells are deprived of DHA, the cells may produce a substitute that is as close as possible in terms of unsaturation and chain length. This replacement fatty acid is often DPA. Alcoholics (and also individuals with depression or attention deficit disorders) are known to have higher levels of DPA. (Pawlosky)

DPA contains one less double bond than DHA. While this may seem insignificant, the efficiency of the receptors in the brain is based on the structure and fluidity of the neuronal membranes. Even the slightest change can cause effects significant enough in the long term to cause noticeable negative results. According to researchers at the National Institute on Alcoholism and Alcohol Abuse, National Institutes of Health, Rockville, MD,, this change is believed to be associated with a loss in nervous system function and may underlie some of the neuropathology associated with alcoholism. (Pawloksky)

The Chicken or the Egg?

In the case of addiction disorders, it is suspected that a deficiency of DHA may be contributing to the alcoholism, and possibly contributing to elevated cortisol levels. The chronic use of alcohol, then seems to exacerbate the deficiency of DHA, and the problem.

DHA, Alcoholism and Depression

Depression often accompanies alcoholism as up to 60% of alcoholics are also depressed. Depression occurs more frequently in alcoholics (58%) than in opiate addicts (32%) or schizophrenics (28%). Researchers have noted serious depression in 70% of patients with prolonged heavy drinking and argued that these depressions were the consequence of the pharmacological effects of alcohol intoxication, withdrawal, and life crises. (Schuckit)

Alcohol is a pro-oxidant that leads to increased lipid peroxidation (a form of free radical damage). A consequence of increased lipid peroxidation may be a decrease in the concentrations of the more highly unsaturated species such as DHA.

Several studies have demonstrated that chronic alcohol intoxication depletes DHA from neuronal membranes which may facilitate development of depressive symptoms. With sobriety, depression may resolve in patients as polyunsaturated fatty acids reaccumulate in the nervous system. This hypothesis predicts that supplementation with DHA and other Omega-3 fatty acids may speed resolution of depressive symptoms in recovering alcoholics.

Depression

Increases in the prevalence of depression in North America during the last 100 years are dramatic and well documented. (Klerman) Since 1900, individuals who were born between 1930 and 1940, have had a higher lifetime risk of depression than those born previous to 1930. Researchers suggest that the increased rates are due to the increased consumption of saturated and Omega-6 fats and the decreased consumption of Omega-3 fats. While the stresses of modern life also contribute to this effect, relative deficiencies in Omega-3 fatty acids intensify vulnerability to depression. The proportion of Omega-6 compared to Omega-3 fats in the diet has markedly increased in the last century because of the few plant species used as sources for Omega-3.

Studies conducted on individuals to lower elevated cholesterol by reducing total dietary fats have revealed an interesting revelation - an increase in suicide, homicide, aggression, hostility and depression. This affect was not seen in individuals who replaced their dietary saturated fats with fish. Researchers postulate that adequate long-chain polyunsaturated fatty acids, particularly DHA, may reduce the development of depression and these other behavior abnormalities. The quantity and balance of (or lack of) dietary polyunsaturated EFAs influences serum lipids and alters the biophysical and biochemical properties of cell membranes.

Several epidemiological studies demonstrate that decreased Omega-3 fatty acid consumption correlates with increasing rates of depression. An article in the *American Journal of Clinical Nutrition* states, "Societies consuming large amounts of fish and Omega-3 fatty acids appear to have lower rates of major depression." According to one study, North American and European populations exhibited cumulative rates of depression ten times that of a Taiwanese population that consumed a diet much higher in fresh fish.

Omega-3 deficiencies may also contribute to depressive symptoms in alcoholism, multiple sclerosis, and postpartum depression.

Dietary polyunsaturated fats and cholesterol are the major determinants of membrane order (or fluidity) in synaptic membranes. Up to 45% of the fatty acids of synaptic membranes are essential fatty acids, mostly DHA and AA. Mammals cannot convert fatty acids between Omega-6 and Omega-3 families. Unsaturated fatty acid composition is a major physiological determinant of the biophysical properties of membranes and DHA seems highly specialized for neuronal membrane function. DHA is replaced by Omega-6 fatty acids in Omega-3 deficiency states, but biophysical properties may not be fully compensated. (Salem)

Researchers at the Department of Psychiatry, University of Sheffield, UK, measured red blood cell membrane fatty acids in a group of depressed patients and compared them to a well matched healthy control group. They found that the severity of the depression correlated negatively with red blood cell membrane levels and with dietary intake of Omega-3s. These findings raise the possibility that depressive symptoms may be alleviated by Omega-3 supplementation, particularly DHA, as it is a crucial component of synaptic cell membranes. (Edwards)

This same research team later found that the depletion of cell membrane Omega-3s, particularly DHA, in depressed individuals also showed evidence of oxidative damage. (Peet) This indicates that Omega-3 supplementation, especially DHA, plus protective antioxidants proven to cross the blood brain barrier such as ginkgo biloba, and alpha lipoic acid would be very beneficial.

Depression and Multiple Sclerosis

The incidence of depression in multiple sclerosis is high and out of proportion to the incidence found in patients with similar disabilities. (Mindon) The depression experienced in these individuals is believed to be due to the depletion of Omega-3 fatty acids in the central nervous system, outside of the plaques. AA and DHA are reduced in normal-appearing white matter, compared with control subjects. (Wilson) Marked reductions of DHA also occur in plasma and adipose stores.

Some researchers suggest that deficient dietary intake of DHA during critical periods of brain development could explain the geographical differences in the incidence of multiple sclerosis. (Bernsohn) It is interesting to note that these geographical differences are similar to differences in rates of major depressions. Treatment with Omega-3 and Omega-6 fatty acids generally results in a mild reduction in relapses and a reversal of depressive symptoms.

Schizophrenia

Schizophrenia is any one of a large group of mental disorders in which the individual loses touch with reality. The individual is unable to think, talk or act normally and the symptoms can range from mild to severe. For years the medical community has accepted the fact that the cause is unknown with little hope for effective treatment. Hopefully, this may be changing due to new research on EFAs and DHA.

Schizophrenic Individuals Have Low DHA Levels

Studies show that schizophrenic individuals demonstrate marked depletion of EFAs, particularly AA and DHA, in red blood cell membranes compared to healthy control subjects. These fatty acid components of cell membrane phospholipids are essential for normal membrane structures, for proper functioning and for normal cell-signalling responses. With inadequate levels of the basic structural material present, how could one expect normal functioning?

Deficient levels of DHA are responsible for a number of physical abnormalities including reduced immunological functions and reduced vasodilatation responses to niacin and histamine suggesting abnormal prostaglandin activity. The normal response to niacin is a flushing sensation which occurs as blood vessels dilate as prostaglandins in the body react. A modest membrane abnormality is likely to produce its most serious consequences in the brain, which requires the delicate, coordinated sequential and parallel activities of millions of neurons. (Horrobin)

Because levels of fatty acids are influenced by diet, medications, and other factors, researchers at the Department of Psychiatry, Medical College of Georgia, in

Augusta, examined cell plasma membrane levels of AA and DHA from 12 schizophrenic patients, eight of whom were drug-naive and in a first episode of psychosis, six bipolar patients, and eight normal control subjects. They found that DHA as well as total Omega-3 EFA contents were significantly lower in schizophrenic patients than from bipolar patients and normal subjects, with no difference between the latter two groups. AA levels did not differ across the groups. (Mahadik)

Symptoms of Apathy and Withdrawal Associated with Depressed DHA Levels

Investigators at Highland Psychiatric Research Group, Craig Dunain Hospital, Inverness, UK, wanted to see if the broad categories of negative and positive symptoms of schizophrenia were linked to specific changes in fatty acids. They found that negative symptoms (apathy and withdrawal) were associated with high levels of saturated fatty acids and low levels of long-chain unsaturates in red blood cell membranes, while the positive symptom patients (either thought disorder or hallucinations and delusions) showed low levels of saturated fatty acids and high levels of long-chain unsaturated fatty acids.

To see if this distribution would be found in patients diagnosed as schizophrenic but without classification of symptoms, they examined levels for fatty acids in plasma and in red blood cell membranes in 68 individuals classified as schizophrenics and 259 normal individuals. Depleted levels were found for AA and DHA in red blood cell membranes from the schizophrenics, but normal fatty acid levels were found in normal membranes. (Glen)

Schizophrenia: A Membrane Lipid Disorder

Research in the last few years provides evidence that schizophrenia involves altered phospholipid-dependent signal transduction (PDST). Researchers at the Clarke Institute of Psychiatry, University of Toronto, Ontario, Canada have demonstrated that an altered response to niacin (vitamin B-3) may reflect disturbances in these signalling processes in this disorder. Niacin induces vasodilation through mechanisms requiring intact PDST. Schizophrenic patients failing to flush with niacin showed significantly reduced levels of AA and DHA. Conversion from non-flushing to flushing during the six-month supplementation period was predicted by an increase in AA levels in red blood cell membranes irrespective of nature of supplementation. (Hudson, Glen)

Clinical, biochemical and genetic evidence now indicates that schizophrenia is a disorder of membrane phospholipid metabolism associated with increased loss of highly polyunsaturated fatty acids from membranes resulting in enhanced activity of phospholipase A2, a rate-limiting enzyme in the synthesis of prostaglandins from AA. Some studies suggest that this is due to a genetic variant. (Hudson)

DHA/AA Abnormalities in Schizophrenia and Dyslexia

In dyslexia, there is evidence for reduced incorporation of DHA and AA into cell membranes, while in schizophrenia, there is evidence for an increased rate of DHA and AA loss from membranes because of enhanced phospholipase A2 activity. The presence of both defects will cause a much greater degree of abnormality than either one alone. It is hypothesized that clinical schizophrenia may occur when both genes are present in the same individual. The dyslexia gene alone will produce dyslexia while the schizophrenia gene alone may produce bipolar or schizoaffective disorders. These proposals could explain the following:

1. The reduced asymmetry of the brain in both schizophrenia and dyslexia;

2. The schizotypal personality characteristics of dyslexics;

3. The increased risks of dyslexia in families with a

schizophrenic proband;

4. The increased risks of bipolar and schizoaffective disorders in families with a schizophrenic proband;

5. The earlier onset and possibly increased severity of both disorders in males since females have a lower requirement for AA and DHA;

6. The absence of selective pressure against schizophrenia since reproduction would be impaired only when the schizophrenic gene coexisted with a dyslexic gene. The schizophrenic gene alone might even lead to improved reproductive performance. (Horrobin)

Omega-3 Reduces Schizophrenia/Tardive Dyskinesia

The Department of Psychiatry, Northern General Hospital, Sheffield, UK, clearly demonstrated dietary increases in Omega-3 fatty acids which modify membrane levels of fatty acids, can have significant effects upon symptoms of schizophrenia and tardive dyskinesia (TD). TD is an abnormal condition involving involuntary repitious movement of the muscles of the face, limbs, and trunk. In a pilot study of Omega-3 supplementation, researchers observed significant improvement in both schizophrenic symptoms and tardive dyskinesia over a six-week period. (Peet)

The World Health Organization schizophrenia study involving eight national centers over two years, showed that the outcome of schizophrenia may be explained by differences in fatty acid intakes, particularly saturated and unsaturated fatty acids. They found highly significant correlations between favorable ratings of course and outcome of schizophrenia and a low intake of total fat and saturated fat and a high intake of unsaturated fatty acids from fish and seafood. The results suggest that the course and outcome of schizophrenia may be influenced through diet. (Christensen)

Alzheimer's Disease/ Memory

Fourteen billion brain cells make up the grey matter of the brain. Each of these cells has connecting arms ending in synapses. These arms transport electrical currents between each brain cell, thus sending messages such as pain and pleasure to the body. When the arms are intact, the communication between brain cells is efficient. If the arms harden due to aging or free radical damage, signals are transmitted slowly and may be altered. If DHA levels are adequate, the connections are more likely to remain functional. If the level drops, the connections become inefficient, which can lead to brain and memory disorders.

As we age, the ability to manufacture long-chain neural fatty acids becomes less efficient. This means that dietary sources become more important. The effects of free radicals on brain tissue becomes more evident as we grow older. With increasing free radical activity comes increasing lipid peroxidation, or rancid fats. Replacing damaged fatty acids in brain membranes may be important in retaining brain function as we get older.

Studies suggest that DHA supplementation may be helpful for the elderly, who require high levels of brain nutrients. The elderly can fall prey to brain deterioration in the form of Alzheimer's disease or senility, both of which can lead to dementia.

Dementia is the term used to describe the loss of mental function. There is new evidence that certain fatty acids may protect against development of dementia and that fatty acids may be useful once dementia has developed.

Dr. Ernst Schaefer, of the Human Nutrition Center on Aging at Tufts University, has discovered that a low level of DHA is a significant risk factor for dementia. Individuals with low blood levels of DHA had almost twice the risk of developing dementia over the next nine years than with those whose blood levels of DHA were high. (Schaefer)

Patients taking DHA as a supplement showed a 65% improvement in dementia symptoms in a study performed at the Gunma National University Medical Department in Japan. The study found that 69% of those whose dementia was due to blood vessel problems improved while on DHA supplements (700-1,400 mg daily). Alzheimer's patients experienced improvement in cooperation, speech, depression, and other psychological symptoms. (Yazawa)

The uptake of fatty acids into the neuronal tissue can be very slow and therefore it may take longer to see changes in the brain and other nervous tissue compared to other aspects in the body such as regulation of blood lipids or other prostaglandin related activities. Do not be discouraged if you do not see results in six or eight months, it may take longer. Many degenerative conditions do not develop "overnight" and cannot be expected to be reversed quickly either.

Alzheimer's Patients Have DHA Deficiency

Recent developments show that the brains of persons who have died from Alzheimer's disease have an EFA deficiency, particularly DHA, a crucial component of synaptic cell membranes.

Dr. Horrobin and colleagues showed that patients with Alzheimer's disease who received essential fatty acids plus antioxidants showed improvement that was consistently better than those not receiving fatty acids. (Corrigan)

Researchers hypothesize that faulty brain cell membranes also resulting from this deficiency may allow a series of reactions allowing the release of a complete sequence of beta amyloid protein into the extracellular space. Beta amyloid protein appears to be the principal active constituent of senile plaques thought to be a probable cause of brain damage resulting in Alzheimer's. (Newman)

DHA and other fatty acids are at high risk for oxidation in the brain contributing to increased degradation of brain phospholipids in Alzheimer's disease. This was demonstrated by researchers revealing significantly decreased brain levels of membrane phospholipids such as phosphatidylethanolamine and phosphatidylinositol, which are made up of DHA, AA and other fatty acids, in Alzheimer's patients compared to control subjects. (Prasad) This suggests that oxidative stress causes degradation of brain phospholipids in Alzheimer's disease suggesting the importance of protective antioxidants such as vitamin E, ginkgo biloba and alpha lipoic acid which have demonstrated helpful to ward off such oxidative damage in the brain and throughout the body.

Phospholipid treatment has also shown some benefit for individuals with memory loss. There are dozens of published human studies examining the effects of phosphatidylserine on improving brain function. In a study of 51 Alzheimer's patients, researchers gave 100 mg of phosphatidylserine three times daily for 12 weeks. Those treated with phosphatidylserine showed improvement on several measures of cognitive function. (Crook)

Another phospholipid, phosphatidylcholine, has also been studied extensively for its effects on mental intelligence and age-related loss of mental function, including Alzheimer's disease. Researchers at Massachusetts Institute of Technology (MIT), Cambridge, demonstrated that if there is inadequate choline supplied through the diet, the body will actually "steal" choline from its own neural membranes to produce the neurotransmitter acetylcholine. (Blusztajn) Researchers speculate that this use of choline from the breakdown of membrane phospholipids (which also contain DHA), might explain the major deficits in long-axon cholinergic nerve terminals seen in Alzheimer's disease and other age-related memory disorders. (Maire)

Inflammatory Disorders

Inflammation is a natural response of the body due to tissue irritation or injury. Inflammatory disorders can result from imbalances created by an over production of series 2 prostaglandins (PGE2). This is often the result of an overconsumption of AA, which comes primarily from animal products.

AA also releases leukotreines, a group of chemical compounds that occur naturally in leukocytes (white blood cells). In excess, leukotreines provoke breast lumps and inflammation such as arthritis, asthma, hives, eczema, dermatitis, psoriasis, allergies, colitis and lupus. Leukotrienes also stimulate the production of mucus, effect cerebral circulation and contribute to various other health problems.

AA and PGE2 also stimulate cell division and proliferation which in the extreme degree can be directly linked to cancer and tumors.

Aspirin and anti-inflammatory steroids work by blocking the production of PGE2. However, aspirin also blocks the production of beneficial "anti-inflammatory" PGE1.

Certain polyunsaturated fatty acids, usually of marine origin, EPA, in particular, may substitute for AA in tissue. Inflammation mediators derived from EPA are much less active biologically, and also can inhibit the harmful behavior of certain leukotrienes (including 4 and 5-series) derived from AA, and reduce platelet-activating factor (PAF).

EPA and DHA Reduce Leukotriene Formation

DHA and EPA are needed by the body to prevent the formation of various harmful inflammatory leukotrienes.

Dietary supplementation leads to incorporation of EPA into membrane phospholipids. Neutrophil (a type of white blood cell) response and the capacity of these cells to adhere to endothelial cells causing inflammation are substantially weakened. This demonstrates that supplemental EPA has anti-inflammatory potential. Clinical trials with EPA and individuals with rheumatoid arthritis, psoriasis, atopic dermatitis, and bronchial asthma have all shown beneficial effects. (Lee)

A number of studies at Cornell University have examined the effects of dietary LNA and fish oil on 4 and 5-series leukotriene formation. The findings demonstrate that EPA and DHA are effective in reducing leukotriene synthesis and that consuming dietary EPA and DHA are 2.5 to 5 times more effective than LNA. They also demonstrated that the degree of effects were dose-related. (Broughton, Whelan) One particular study demonstrated a 76% reduction in leukotriene synthesis in response to an inflammatory stimulus following a dietary regimen containing fish oil. (Broughton)

Animal studies at Cornell University showed the following effects on prostaglandin synthesis following increased dietary intake of Omega-3 fatty acids relative to Omega-6 fatty acids:

■ Sulfidopeptide leukotrienes (SP-LT) (LTC4 and LTE4) was progressively reduced by 76%.

■ 6-keto-prostaglandin F1 alpha decreased by 81%.

■ Prostaglandin E2 synthesis decreased by 44%. (Broughton)

High-Fish Diet Helpful for Heart Health Also Helpful for Inflammatory Conditions

Researchers at the Human Nutrition Research Center on Aging, Tufts University, Boston, Massachusetts investigated the immunologic effects of the diets recommended by the National Cholesterol Education Panel. Reductions in dietary fat, saturated fat, and cholesterol have been recommended to reduce the risk of heart disease in our society.

Twenty-two individuals over 40 years of age were first given a diet for six weeks which is similar to the current American diet:

14.1% of calories as saturated fatty acids,
14.5% monounsaturated fatty acids,
6.1% Omega-6, 0.8% Omega-3,
147 mg. cholesterol (per 1,000 calories).

They were divided into two groups. Both groups were to receive the base diet which was low-fat, low-cholesterol, and high-polyunsaturated fatty acid (PUFA) recommended by the National Cholesterol Education Panel. One group received a low-fat, low-fish version, the other group received a low-fat, high-fish version for 24 weeks:

Low-fat, low-fish group:

4.0-4.5% saturated,
10.8-11.6% monounsaturated,
10.3-10.5% PUFA, low in fish-derived Omega-3 [0.13% or 0.27 grams/day EPA and DHA] - [33 grams fish/day].
45-61 mg. cholesterol, per 1,000 calories.

Low-fat, high-fish group:

Same diet except that it was enriched in fish-derived Omega-3s (1.23 grams/day EPA and DHA) and 121-188 grams fish/day.

The low-fat, low-fish diet increased inflammatory responses through various means (increased tumor necrosis factor (TNF), etc.) and had no beneficial effect on delayed-type hypersensitivity skin response, or IL-6, GM-CSF, or prostaglandin type-2 (PGE2) production.

In contrast, the low-fat, high-fish diet significantly improved parameters helpful for inflammatory skin conditions. It decreased the percentage of helper T cells whereas the percentage of suppressor T cells increased. Responses to delayed-type hypersensitivity skin response as well as the production of cytokines IL-1 beta, TNF, and IL-6 were significantly reduced after the consumption of the low-fat, high-fish diet.

The data further demonstrates that the high-fish diet significantly decreases various parameters of the hyper-immune response associated with allergies, asthma, arthritis, etc. The researchers concluded that such alterations may be beneficial for the prevention and treatment of atherosclerotic and inflammatory diseases. (Meydani)

DHA More Beneficial Than EPA in Inflammatory Conditions

Most studies conducted on fish oils use a combination of EPA and DHA. This particular animal study examined them separately to determine if they have different effects or to determine if one was more effective than the other.

To examine their effects on acute inflammation, this study compared the effects of a diet containing either 1% DHA, EPA, or corn oil (placebo). The dietary treatment with DHA or EPA elevated the Omega-3 fatty acids as expected in the spleen and phospholipids, associated with a reduction in AA levels. The degree of inflammation was quantified by measuring the four parameters including:

1. Edema (inflammation),

2. Polymorphonuclear cell infiltration as myeloperoxidase activity (MPO) at the site of inflammation,

3. Prostaglandin E2 (PGE2) concentrations,

4. Leukotriene B4 (LTB4) concentrations

The addition of DHA to the diet reduced significantly the edema formation and the MPO activity 24 hours after challenge. Both DHA and EPA significantly reduced the harmful PGE2 and LTB4 levels compared with animals fed corn oil. This result suggests that DHA rather than EPA may be more useful in the adjuvant treatment of diseases where acute inflammatory processes play a role. (Raederstorff)

Researchers in France also demonstrated that DHA is more effective than EPA in inhibiting a variety of inflammatory mediators, but the two compounds in association had an additive effect. This inhibitory effect may explain the protective and beneficial effects of Omega-3s on diseases involving chronic inflammatory reaction. (Khalfoun)

Allergies/Asthma

Researchers in Japan report that there has been a significant rise in the rate of allergies among Japanese babies, estimating that one-third of the infants born in Japan are now diagnosed with allergic disorders. This is attributed to the Westernized changes in the Japanese diet, a reduced intake of Omega-3, combined with excessive Omega-6 intake. This leads to the overproduction of irritating Omega-6 prostaglandins. (Okuyama)

Mother's DHA Intake Prevents Allergies in Children

Pregnant women with allergies should pay close attention to secure that their dietary intake of essential fatty acids (EPA and DHA in particular) is adequate to insure that they do not pass on allergic sensitivities to their offspring. Swedish researchers found an inverse relationship between serum levels of phospholipid fatty acids in allergic mothers (but not non-allergic mothers) and their babies in relation to allergic disease. The maternal dihomo-gamma-linolenic acid (DHGLA) levels correlated with the umbilical cord serum levels of AA and DHA. The findings indicate that there is an abnormal metabolism relationship between some of the long-chain polyunsaturated fatty acids in allergic mothers, affecting their infants. Furthermore, the findings suggest an association between the fatty acid composition in maternal serum and the appearance of allergic disease in their children during the first six years of life.

The proportions of various long-chain polyunsaturated fatty acids were altered in the serum phospholipids of allergic pregnant mothers and in mothers whose babies developed allergic disease over the first six years of life, indicating that atopy is associated with a disturbed fatty acid metabolism. (Yu)

Other researchers studying the relationship between leukotrienes, fish-oil, and asthma at the Department of Rheumatology, Brigham and Women's Hospital, Boston, Massachusetts have found similar results. Their studies suggest that leukotrienes which have been metabolized from AA released from membrane phospholipids during cell activation may play a significant role in a variety of inflammatory disorders including the pathophysiology of chronic allergic asthma. They explain that EPA and DHA limit leukotriene synthesis and biological activities by substituting substrate fatty acids as alternatives to AA. Both EPA and DHA inhibit the conversion of AA to prostanoid metabolites and reduce the production of PAF. (Arm)

PAF is a lipid mediator released by many kinds of cells which exerts its effects on blood cells and on cells of the bronchial wall both directly and indirectly. It causes airway edema, eosinophil accumulation in the airway wall, and bronchial hyperresponsiveness. PAF has potent activity as a chemotactic agent and as an activator of eosinophils, which are prominent white blood cells in asthmatic airways, through the activation of specific surface receptors. The interaction between PAF and eosinophils may be crucial in the pathogenesis of bronchial hyperresponsiveness and inflammation in asthma.

French researchers examined the effect of a fish oil diet on 12 asthmatic patients in a one-year double-blind study. They found a positive effect on forced expiratory volume after nine months of treatment favoring the use of EPA and DHA. (Dry)

Fish Oils Reduce Airway Response to Inhaled Allergen in Bronchial Asthma

Researchers at the Department of Allergy and Allied Respiratory Disorders at United Medical School, Guy's Hospital in London, examined the effects of dietary supplementation with fish oil lipids on the airways response in asthmatics.

Nine asthmatics received fish oil capsules containing a total of 3.2 g EPA and 2.2 g DHA a day and eight controls received identical capsules containing olive oil, for 10 weeks in a double-blind fashion. When the individuals were challenged with an allergen, the asthmatic response was significantly reduced (from two to seven hours) after allergen challenge following dietary supplementation with EPA/DHA but not with placebo. (Arm)

Another group of researchers at the University of Wyoming in Laramie, found that the reduced asthma symptoms due to Omega-3 fatty acid ingestion are related to 5-series leukotriene production. Omega-3s in ratios to Omega-6s of 0.1 to 1 and 0.5 to 1 were ingested sequentially for one month each. Patient respiratory indexes were assessed after each treatment. Forced vital capacity, forced expiratory volume, peak expiratory flow, and forced expiratory flow 25-75% were measured along with weekly 24-hour urinary leukotriene concentrations.

With low Omega-3 ingestion, respiratory distress increased. With high Omega-3 ingestion, alterations in urinary 5-series leukotriene excretion predicted treatment efficacy. Elevated Omega-3 ingestion resulted in improvement in over 40% of the test subjects. Five-series leukotriene excretion with high Omega-3s ingestion was significantly greater for those who improved compared to those who did not. (Broughton)

Skin Disorders

T-cell activation and cytokine production play an important role in several chronic inflammatory diseases. Omega-3s exert beneficial effects on the clinical state of some of these diseases.

Researchers in Norway examined the effect of Omega-3 supplementation on T-cell proliferation, expression of CD25 (interleukin-2 receptor), secretion of interleukin-2, interleukin-6 and tumor necrosis factor from T-cells from patients with psoriasis and atopic dermatitis.

Researchers have demonstrated the effectiveness of Omega-3 fatty acids (supplementing capsules containing six grams EPA and DHA daily) to decrease interleukin-2 receptors on lymphocytes in patients with atopic and psoriatic dermatitis. (Soyland)

Psoriasis

Psoriasis may improve during dietary supplementation with fish oil containing Omega-3 fatty acids including EPA. In one study, 17 psoriatic patients were treated with capsules containing a combination of Omega-3 and Omega-6 fatty acids. After four months, excellent improvement was observed in two patients, moderate improvement in eight, mild improvement in four, and no improvement in three patients. These results may indicate that a combination of Omega-3 and Omega-6 fatty acids is useful for the treatment of psoriasis. (Kragballe)

Researchers at the Department of Dermatology, University Central Hospital, Helsinki, Finland examined the effects of DHA and EPA in patients with psoriasis and psoriatic arthritis. The results were very encouraging.

Eighty patients with chronic, stable psoriasis, 34 of whom also had psoriatic arthritis, were treated with 1,122 mg/day EPA and 756 mg/day DHA. Before the study and after four and eight weeks of treatment a Psoriatic Association Scoring Index (PASI) score was assessed.

	Before treatment	After 4 weeks	After 8 weeks
Mean PASI score	3.56	1.98	1.24

The degree of pruritus (itching) decreased most rapidly, followed by scaling and induration of the plaques, and erythema was most persistent. At the end of the trial, seven patients were completely healed and in 13 other patients more than 75% healing was observed, but in 14 patients the result was poor.

The majority of patients with psoriatic arthritis reported a subjective improvement in joint pain during the study. It was concluded that EPA and DHA may be useful for the treatment of psoriasis and psoriatic arthritis and may provide an important adjuvant to standard therapy of both conditions. (Lassus)

Fish Oils Reduce Over-Sensitivity to UV Radiation

Ciniforme is a troublesome and scarring photosensitivity disorder for which no satisfactory treatment is currently available. Studies suggest dietary fish oil rich in Omega-3s may be helpful as it increases the resistance to ultraviolet-induced redness, swelling and rash provocation.

Three Caucasian boys with the condition were given capsules containing 1.65 grams EPA and 1.1 grams DHA daily. Phototesting was performed at baseline and after three months supplementation. At baseline, low erythemal thresholds were seen to monochromated UVA in all three boys. Following fish oil, all the boys showed reduced sensitivity to UVA and one also showed reduced sensitivity to UVB. Provocation challenge revealed a reduced response in all three children. Clinically, these changes were accompanied by pronounced overall improvement in one child, mild improvement in the second child, but no improvement in the third. (Rhodes)

EPA/DHA May Benefit Raynaud's Phenomenon

Researchers at the Division of Rheumatology at Albany Medical College, New York investigated the effects of fish-oil supplementation in patients with Raynaud's phenomenon and rheumatic disease. Raynaud's involves sporadic attacks of blood flow interruption (ischemia) of the fingers, toes, ears, and nose. It is caused by exposure to cold or by emotional stimulation. The individual experiences numbness, tingling, burning and pain. The attacks are often linked to other health conditions such as rheumatoid arthritis, and also drug poisoning.

Omega-3 fatty acids could benefit patients with Raynaud's phenomenon because, among other effects, these fatty acids induce a favorable vascular response to ischemia. Thirty-two patients with primary or secondary Raynaud's phenomenon were randomly assigned to olive-oil placebo or fish-oil groups. Patients ingested 12 fish oil capsules daily containing a total of 3.96 gEPA and 2.64 g DHA or 12 olive oil capsules.

In the fish-oil group, the median time interval before the onset of Raynaud's phenomenon increased from 31.3 minutes baseline to 46.5 minutes at six weeks. Patients with primary Raynaud's phenomenon ingesting fish oil had the greatest increase in the time interval before the onset of the condition. Five of 11 patients (45.5%) with primary Raynaud's phenomenon ingesting fish oil in whom the phenomenon was induced at baseline could not be induced to develop Raynaud's at the six- or 12-week visit compared with one of nine patients (11%) with primary Raynaud's ingesting olive oil. The average digital systolic pressures were higher in the patients with primary Raynaud's phenomenon ingesting fish oil than in patients with primary Raynaud's ingesting olive oil in the 10 degrees C water bath.

The researchers conclude that the ingestion of fish oil improves tolerance to cold exposure and delays the onset of vasospasm in patients with primary, but not secondary Raynaud's phenomenon. These improvements are associated with significantly increased digital systolic blood pressures in cold temperatures. (DiGiacomo)

Arthritis

The degenerative changes associated with arthritis, effect the joints, tendons, ligaments and muscles. Inflammation, stiffness and pain are among the signals that something is wrong.

EFAs are needed to produce secretions to lubricate the joints. EPA and DHA are needed to suppress the formation of leukotrienes and inflammatory prostaglandins as well as support the production of anti-inflammatory prostaglandins.

Leukotreines are formed from AA, largely by white blood cells called neutrophils, found in large numbers at the site of inflammation in individuals with rheumatoid arthritis. DHA supplementation has demonstrated to reduce the formation of leukotrienes as much as 76%. (Broughton)

DHA Reduces Inflammatory Response

Omega-3 fatty acid supplementation reduces the response of neutrophils (a type of white blood cell) and their generation of inflammatory leukotriene (LT B4).

Researchers at the Department of Medicine, Harvard Medical School in Boston, wanted to investigate this aspect of Omega-3s. Neutrophils and monocytes from eight healthy individuals were examined before and after they were given dietary supplementation of 9.4 grams EPA and 5 grams DHA daily.

The neutrophil response to LTB4 decreased by 69% after three weeks and by 93% after 10 weeks from pre-diet values. Other damaging responses by neutrophils stimulated by LTB4 decreased by 71% after three weeks and by 90% after 10 weeks from pre-diet values. Each response correlated closely and negatively with the EPA content of the neutrophil phosphatidylinositol pool respectively. (Sperling)

Fish Oil Reduces Joint Swelling and Morning Stiffness

A randomized, double-blind, placebo-controlled crossover study in the Netherlands demonstrated the beneficial effects of EPA and DHA in individuals with rheumatoid arthritis. The individuals added EPA/DHA supplementation to their normal existing treatments with non-steroidal anti-inflammatory drugs and with disease-modifying drugs. The results favored fish oil supplementation over the placebo. The most significant improvements were noted in reduced joint swelling and reduced duration of early morning stiffness. Other clinical indices improved but did not reach statistical significance.

During fish oil supplementation relative amounts of EPA/DHA in the plasma cholesterol ester and neutrophil membrane phospholipid fractions increased, mainly at the expense of the Omega-6 fatty acids. The mean neutrophil leukotriene B4 production in vitro showed a reduction after 12 weeks of fish oil supplementation. Leukotriene B5 production, which could not be detected either in the control or in the placebo period, rose to substantial quantities during fish oil treatment. The researchers concluded that dietary fish oil supplementation is effective in suppressing clinical symptoms of rheumatoid arthritis. (van der Tempel)

In another controlled, double-blind, clinical trial researchers tested the effect of dietary Omega-3 fatty acid supplementation with and without naproxen (a non-steroidal anti-inflammatory agent) and placebo, respectively, in 67 patients with active rheumatoid arthritis. The results also indicated significant improvement in most variables measured in the Omega-3 fatty acid group compared to the placebo group with morning stiffness significantly less pronounced. (Kjeldsen-Kragh)

EPA/DHA Reduces Need for Anti-inflammatories

Researchers in Scotland investigated the effects of fish oil supplementation on non-steroidal anti-inflammatory drug (NSAID) requirement in patients with rheumatoid arthritis in a double-blind, placebo-controlled study.

The anti-inflammatory properties of EPA/DHA could encourage one to expect the requirement for NSAIDs in patients with rheumatoid arthritis to be reduced. DHA and EPA reduces the formation of the inflammatory prostaglandins (3-series) and leukotrienes (5-series).

Sixty-four patients with stable rheumatoid arthritis requiring NSAID therapy were studied. Patients received either 10 fish oil capsules (each containing 171 mg. EPA and 114 mg. DHA) or placebo capsules per day for 12 months. All then received placebo capsules for a further three months. NSAID requirement at the beginning of the study for each patient was assigned as 100%. Patients were instructed to slowly reduce their NSAID dosage providing there was no worsening of their symptoms. Clinical and laboratory parameters of rheumatoid arthritis activity were also measured.

There was a significant reduction in NSAID usage in patients on DHA/EPA when compared with placebo.

	DHA/EPA	Placebo
Month 3	71.1%	89.7%
Month 12	40.6%	84.1%
Month 15	44.7%	85.8%

(The test group changed back to placebo after month 12 for the last three months of the trial)

The DHA/EPA patients were able to significantly reduce their NSAID requirement without experiencing any deterioration in the clinical and laboratory parameters of rheumatoid arthritis activity. (Lau)

DHA Reduces Atherogenic and Inflammatory Proteins

A number of studies by researchers at Harvard Medical School in Boston, have helped reveal the specific and complex mechanisms by which DHA can benefit atherogenesis and inflammation. These studies also demonstrate that DHA, and not EPA, is more beneficial in these investigations concerning reduced inflammation and atherogenesis.

Induction in endothelial cells of adhesion molecules for circulating leukocytes and of inflammatory mediators by cytokines is thought to contribute to the early phases of atherogenesis and inflammation.

The researchers report that DHA incorporation into cellular lipids decreases cytokine-induced endothelial leukocyte adhesion molecules, secretion of inflammatory mediators, and leukocyte adhesion to cultured endothelial cells. DHA, but not EPA, decreased in a dose - and time-dependent fashion the expression of vascular cell adhesion molecule 1 induced by interleukin (IL)-1, tumor necrosis factor (TNF), IL-4, or bacterial lipopolysaccharide.

DHA also limited cytokine-stimulated endothelial cell expression of E-selectin and intercellular adhesion molecule 1 and the secretion of IL-6 and IL-8 into the medium but not the surface expression of constitutive surface molecules. DHA treatment also reduced the adhesion of human monocytes and of monocytic U937 cells to cytokine-stimulated endothelial cells. These properties of DHA contribute to its antiatherogenic and anti-inflammatory effects.

They also showed the importance of long-term supplementation and showed that higher intake resulted in stronger protective benefits. (De Caterina)

Crohn's Disease/ Ulcerative Colitis

Crohn's disease is characterized by diarrhea, abdominal pain, fever, weight loss, and weakness due to inflammation of the small intestine, and sometimes, large intestine. While the ailment tends to be chronic, patients may have remissions between flare-tips. Research shows that bowel tissue in individuals with Crohn's disease contains abnormally high levels of inflammatory prostaglandins from AA. (Shoda)

Doctors in Italy gave fish-oil capsules containing 2.7 grams of EPA and DHA every day for one year to 39 Crohn's patients who had been in remission for about eight months. A control group of 39 patients received placebo capsules. After a year, 23 of the patients receiving fish oils were still in remission. Only 11 of the 39 control patients stayed in remission.

In the patients taking the fish-oil who remained in remission, red blood cell levels of AA dropped, while levels of EPA and DHA increased greatly. Laboratory tests indicated that inflammation decreased in the fish-oil patients, but increased in the control patients.

Ulcerative colitis is another chronic inflammatory disease, usually of the large intestine, in which ulceration and erosion of the bowel tissue cause severe diarrhea, loss of blood, weakness and weight loss. As in Crohn's disease, bowel tissue shows high levels of AA-produced prostaglandins. The worse the symptoms, the higher the levels of these prostaglandins.

In one study conducted through the Washington University School of Medicine, 24 patients with active ulcerative colitis took fish-oil capsules containing 5.4 grams of EPA and DHA for four months. This caused a substantial reduction in inflammatory prostaglandins, bowel tissue healing and reduced rectal bleeding. They also gained weight - a good sign in their case. In addition, seven patients who were also receiving an anti-inflammatory steriod were able to reduce their dosages in half. (Stenson)

Coronary Artery Disease

The low incidence of acute myocardial infarction they verified within the Greenland Eskimos suggested that a high dietary Omega-3 intake due to marine food might protect against coronary heart disease. They showed that the Eskimos had a beneficial lipid pattern and that their balance between clotting factors was shifted towards an anti-thrombotic state.

Experimental studies support the hypothesis that dietary DHA and EPA intake may play a leading role in primary or secondary prevention of coronary heart disease.

High levels of DHA and EPA are recognized to have the following beneficial effects on heart health:

1. Lower plasma triglyceride levels

2. Lower harmful VLDL and LDL levels

3. Elevate protective HDL levels

4. Lower total serum cholesterol

5. Reduce harmful clotting factors derived from elevated AA (prostaglandins and leukotrienes), which also cause inflammatory problems and disrupt healthy immune cell function

6. Reduce hypertension

DHA Reduces Elevated Triglycerides

Researchers at the Chicago Center for Clinical Research evaluated the effects of supplementation with 1.25 g or 2.5 g of DHA, in the absence of EPA, on serum lipids and lipoproteins in persons with combined hyperlipidemia (elevated blood lipids). They were divided into three groups 1.) placebo, 2.) receiving 1.25 g/day DHA 3.) receiving 2.5 g/day DHA for six weeks of treatment.

The DHA content of plasma phospholipids increased dramatically (two to three fold) in a dose-dependent manner. Significant changes were observed in serum triglycerides (17-21% reduction) and HDL (6% increase) which were of similar magnitude in both DHA groups.

Dietary DHA, in the absence of EPA, can affect cholesterol and triglyceride levels in patients with combined hyperlipidemia. The desirable triglyceride and HDL-cholesterol changes were present at a dose which did not significantly increased non-HDL-cholesterol or LDL-cholesterol. These findings suggest that supplementation with 1.25 grams of DHA per day, may be an effective tool to aid in the management of hypertriglyceridemia. (Davidson)

Researchers at Western Human Nutrition Research Center, in San Francisco, California, investigated the effect of dietary DHA on plasma lipoproteins and tissue fatty acid composition in humans.

Normal, healthy male volunteers were fed diets containing 6 grams per day of DHA for 90 days. The stabilization (low-DHA) diet contained less than 50 mg per day of DHA. A control group remained on the low-DHA diet for the duration of the study (120 days).

	LDL-C	Apo-E	Trig	HDL
Control				
Day 30:	NC	8.46	NC	NC
Day 75:	NC	8.59	NC	NC
Low DHA diet				
Day 30:	NC	NC	76.67 mg/dL	34.83
Day 75:	NC	NC	63.83 mg/dL	37.83
High DHA diet				
Day 30:		7.06 mg/dL		
Day 90:		12.01 mg/dL		

The percentage of plasma DHA rose from 1.83 to 8.12 after 90 days on the high-DHA diet. Although these volunteers were eating a diet free of EPA, plasma EPA levels rose from 0.38 to 3.39 (wt%) after consuming the high-DHA diet. The fatty acid composition of plasma lipid fractions (cholesterol, triglycerides, and phospholipids), showed marked similarity in the enrichment of DHA, about 10%, after the subjects consumed the high-DHA diet. The DHA content of these plasma lipid fractions varied from less than 1% triglycerides to 3.5% at baseline, study day 30. EPA also increased in all plasma lipid fractions after the subjects consumed the high-DHA diet.

There were no changes in the plasma DHA or EPA levels in the control group. Consumption of DHA also caused an increase in adipose tissue levels of DHA, but not EPA. Thus, dietary DHA will lower plasma triglycerides without EPA, and DHA is converted to EPA in significant amounts. Dietary DHA appears to enhance apo-E synthesis in the liver. (Nelson)

DHA More Effective Than EPA

Researchers determined that individuals consuming either four grams EPA or DHA added to a high fat meal had very different triglyceride levels after the meal. EPA recipients lowered their triglyceride levels by 19% compared to controls while DHA recipients lowered their triglyceride levels by 49%. (Hansen)

To compare the effects of EPA and DHA on serum lipids, apolipoproteins, and serum phospholipid fatty acids in humans, researchers in Norway conducted a double-blind, placebo-controlled study. Healthy nonsmoking men (36-56 years old) were randomly assigned to supplementation with 3.8 grams EPA per day, 3.6 grams DHA per day, or a placebo for seven weeks.

Serum triglycerides decreased 26% in the DHA group and 21% in the EPA group compared with the placebo group. Although not significant, decreases in serum triglycerides were consistently greater in the DHA group. HDL cholesterol increased in the DHA group. In the EPA group, serum total cholesterol decreased and apolipoprotein A-I decreased.

In the DHA group, serum phospholipid DHA increased by 69% and EPA increased by 29%, indicating conversion of DHA to EPA. In the EPA group, serum phospholipid EPA increased by 297% whereas DHA decreased by 15%, suggesting that EPA cannot be converted to DHA in humans.

The serum phospholipid ratio of Omega-3 to Omega-6 fatty acids increased in both groups, whereas the relative changes in Omega-6 fatty acids suggested possible alterations in liver desaturation activity in the DHA group. We conclude that both DHA and EPA decrease serum triglycerides, but have differential effects on lipoprotein and fatty acid metabolism in humans. (Grimsgaard)

EPA/DHA Inhibit Platelet Aggregation

Platelets are tiny disc-shaped cells that flow through the blood. One of their primary responsibilities in the body is as a clotting agent. Clotting, or coagulation of the blood. is necessary to prevent blood loss when a vessel is damaged. The platelets adhere to the injury and initiate clot formation.

Platelets can also initiate and accelerate atherosclerosis. When they attach abnormally to an artery wall they stimulate cell division in the smooth muscle cells which narrows the passageway for blood flow. When they adhere to a plaque, they can form a clot which can block off blood supply completely causing a heart attack or stroke.

Researchers at the Department of Pharmacology, University of Illinois in Chicago, found that platelet aggregation was inhibited after supplementation of DHA or EPA. This was demonstrated by the finding that neither fatty acid significantly inhibited prostaglandin-independent aggregation. Remember series-2 prostaglandins, PGE2, promote platelet aggregation.

Enrichment with DHA or EPA resulted in inhibition. They determined that prostaglandin H2 receptor affinity decreased 4.8 fold following DHA incorporation. (Swann)

German researchers investigated the effects of daily

dietary supplementation for six weeks with either 4.5 grams EPA and 3.35 grams DHA (group I) or 3.5 grams EPA and 6.4 grams DHA (group II) on platelet prostaglandin responsiveness.

In group I the response was minimal. In group II, characterized by a high intake of DHA, a considerable reduction in platelet prostaglandin responsiveness was found. These results suggest different effects of EPA and DHA on platelets, which DHA having a stronger effect than EPA. (Scheurlen)

You may wonder about the increased risk of nose-bleeds and bruising if DHA decreases the clotting ability of the blood. If these are a concern increase your intake of antioxidants. I recommend Alpha Lipoic Acid, between 100-200 mg. two to three times daily to help strengthen the capillary walls.

EPA/DHA Most Effective Together to Reduce Clotting

The effects of a fish-enriched diet or dietary supplements consisting of either fish oil or DHA-oil on platelet aggregation and hemostatic factors were studied in healthy male students. After an experimental period of 15 weeks, they determined that fish oil containing both EPA and DHA was more effective than DHA alone.

In the DHA-oil group there was a slight, statistically insignificant, increase of platelet aggregation which correlated significantly with the decrease of plasma triglycerides. Platelet aggregation measured four hours after a standardized fat meal was lower than in the fasting state and this decrease correlated with the increase of plasma triglycerides. These results show that a fish diet and fish oil, inhibited in vitro platelet aggregation. (Agren)

Fish Oils Reduce Hypertension

Hypertension is not only a major risk factor for cardiovascular disease, it may be the most prevalent chronic condition in Western society. In spite of the availability of hypertensive medications, a high proportion of individuals still have poorly controlled blood pressure.

There is a large body of evidence showing that increased consumption of DHA and EPA in fish oils can lower blood pressure. (Howe, 1995) It is believed that this is accomplished through a shift in eicosanoid production away from the series-2 prostaglandins derived from AA, to the series-3 prostaglandins derived from EPA, combined with the competitive inhibition of AA metabolism by both EPA and DHA. (Kinsella) This could result in reduced levels of potent vasoconstrictors such as thromboxane A2.

These effects of fish oils may be enhanced by also reducing sodium intake. In one study with elderly individuals, fish oil combined with a low-sodium diet caused blood pressure to fall by 7/5 mmHg, which is considerably more than either treatment alone. (Howe, 1997)

Fish May Protect Smokers From Heart Disease

Cigarette smokers are also at a greatly increased risk for lung diseases, including lung cancer. The number of lung cancer-related deaths in Japan is far less than that here in the United States, yet smoking is almost twice as common among Japanese men as in American men. This protection from the harmful effects of smoking may be due to their low-fat diet which has a high intake of fatty fish and green tea rich in antioxidants.

Studies from the Honolulu Heart Program have shown the protective effect of fish consumption on the lungs of smokers. This particular study was derived from the 1965 to 1968 program in Oahu, Hawaii where 8,006 Japanese-American men, aged 45 to 68 years, were followed up for the development of heart disease and stroke. A separate 23-year long study was done to determine if fish intake would protect cigarette smokers from an increase in risk of coronary heart disease. The individuals for this particular study included past and current smokers.

They found no significant association between fish intake and the risk of coronary heart disease among the current, past, or never smokers. However, among the 3,310 current smokers (which constituted 44% of the entire group), those who had a low fish intake had an increasing incidence of coronary heart disease depending upon the number of cigarettes smoked per day. The incidence of coronary heart disease was more than twice as high in men who smoked over 30 cigarettes per day, compared to those who smoked less than 20 per day. The death rates among those who smoked heavily (more than 30 cigarettes a day) were three times higher, while those who smoked 20-30 cigarettes a day had a 50% higher death rate compared to those who smoked less than 20 per day.

Among the current smokers who had high fish intake, the number of cigarettes smoked per day had no effect on the incidence of heart disease and the death rates from heart problems.

As seen with lung diseases, the consumption of fish seems to protect smokers from the harmful effects of smoking on the heart and blood vessels. It has been suggested that fish oils alter some metabolic processes that result in beneficial effects including more dilatation of blood vessels, less platelet adhesiveness, reduced inflammatory response to the injury caused by smoking, lower triglyceride and fibrinogen levels, and lower blood pressure (especially in patients who have mildly elevated blood pressure).

The above study showed that the highest incidence rate of coronary heart disease was found among the heavy smoking group with low fish intake. Heavy smokers (more than 30 cigarettes per day) with a high fish intake had half the risk of dying from heart disease compared to heavy smokers with a low fish intake.

The best solution is to quit smoking altogether since nonsmokers have the lowest risk of dying from heart disease, regardless of their fish intake. (Rodriguez)

Diabetes

Diabetes is the third leading cause of death in the U.S. It is responsible for serious suffering in the form of blindness, nerve damage, heart disease, gangrene, and loss of limbs. About half of those with coronary artery disease and three-forths of those suffering strokes developed their circulatory problems prematurely as a result of diabetes.

Diabetes involves the body's inability to properly metabolize food into energy. The result is a build-up of blood sugar that causes a number of serious problems. Even through self-regulation of blood sugar through home glucose tests, medication and insulin, normal glucose levels for a diabetic are almost twice the normal range.

Insulin-dependent diabetes mellitus (Type I) normally results when the body does not produce enough insulin. This is normally the result of damage to the beta cells of the pancreas. This form of diabetes usually, but not always, begins in childhood; thus it is often referred to as juvenile diabetes. This type of diabetes makes up about 10 to 15% of all individuals with diabetes.

Non-insulin-dependent diabetes (Type II) accounts for 85 to 90% of diabetes cases and is usually associated with age and/or obesity. It is sometimes called adult-onset diabetes. Individuals are able to manufacture plenty of insulin, but the body is not efficient in using it to burn ingested carbohydrates resulting in elevated glucose levels.

This condition, referred to as insulin resistance, is caused by being overweight, improper food choices, lack of exercise, stress caused by illness or injury, and certain medications. This form is caused by insulin resistance of cells or the inability of insulin receptors to utilize insulin efficiently. Usually, diet and oral medication can keep blood sugar levels near normal, but insulin is generally of no value. Type II diabetics generally have high levels of insulin in the blood, but it is ineffective because of the insulin resistance of the tissues.

All hormones, including insulin and glucagon, control cells by stimulating production of prostaglandins. These are made of EFAs, so if a deficiency exists (which is common among diabetics), their production is disrupted which can intensify adult-onset diabetes even if adequate insulin in produced.

Diabetics Have Depressed Levels of LNA, EPA and DHA

Omega-3 fatty acids affect the body's ability to respond to insulin. Omega-3s are needed for the cell membranes in order for the tissues to respond to insulin. Higher levels of Omega-6 diminished the tissue's response to insulin.

The high level of glucose in diabetics interferes with their ability to convert LNA into DHA. This is significant for a number of reasons:

1. DHA (and EPA) improves the function of insulin receptors, which helps lower glucose levels.

2. DHA helps lower glucose levels as it increases blood insulin concentrations. (Hamazaki)

3. DHA is an important structural component of the retina. Degeneration of the retina is a common cause of blindness in diabetics.

4. DHA helps regulate blood lipid levels, a risk factor for the development of atherosclerosis. Over 26 published trials have demonstrated the benefits of adding fish oils EPA and DHA to lower triglyceride levels among diabetes, both insulin-dependent (IDDM) and non-insulin dependent (NIDDM). (Friedberg) *Note: Studies were conducted on a dosage level of 1 gram DHA and 1 gram EPA per day.*

EPA/DHA Beneficial for Lipid Levels in Type 2 Diabetes

Elevated triglycerols, VLDL, and HDL, are among the common problems increasing the risk of cardiovascular complications in NIDDM. Japanese researchers investigated the effects of EPA and DHA on 21 NIDDM patients. The patients were treated for 28 days with 1.7 grams EPA and 1.15 grams DHA per day. After the DHA/EPA treatment there was a strong decrease in triglycerides and VLDL serum levels, accompanied by a significant increase in HDL. (Haban)

Other studies on the use of fish oil have also demonstrated that it lowers triglyceride levels effectively by almost 30% and that it has no adverse affects on glucose in diabetics. (Friedberg)

EPA/DHA Increases Glucose Disposal

To determine the impact of fish-oil supplementation on glucose and lipid metabolism in patients with impaired glucose tolerance, eight obese subjects with impaired glucose tolerance (average age 50.3) were given 3.8 grams EPA and 2.5 grams DHA in addition to their regular diet for two weeks. Glucose disposal rate increased after fish oil supplementation, whereas no change was seen without fish oil. (Fasching)

Neuropathy

Neuropathy (painful swelling and destruction of the nerve and nerve endings) is a common problem that accompanies diabetes. Individuals feel numbness, pain and "pins and needles" sensations in their hands and feet with this type of nerve damage. It can make simple tasks like putting socks on or walking down stairs extremely difficult.

Neuropathy can also be the result of drug poisoning (common among individuals taking powerful medications for HIV), lead poisoning, and alcoholism. Other causes of neuropathies include viral infection and autoimmune disorders such as arthritis, lupus, and periarteritis (a disease of the small arteries that can lead to hypertension, heart attack, muscle weakness, skin ulcerations, and gangrene, which can lead to amputation). These conditions are accompanied by elevated levels of free radicals and depressed levels of antioxidants.

Oxidative stress seems greater in neurological tissues because of the sensitivity of their phospholipid structure, and their constant high rate of energy and oxygen consumption. Free radicals may be largely responsible for neurodegenerative diseases.

Individuals with polyneuropathy (meaning there is damage to several nerves in the body) have been found to have lower levels of the potent antioxidant alpha lipoic acid in their body. Studies have shown that alpha lipoic acid is an effective neuroprotective agent.

Diabetics also commonly have depressed levels of Omega-3 fatty acids, LNA, EPA and DHA.

Impaired conversion of LA to GLA has been demonstrated in diabetics. Experts in the field of fatty acid research, such as Dr. David Horrobin, have suspected that this impairment could lead to defective nerve function because metabolites of GLA are known to be important in nerve membrane structure, nerve blood flow, and nerve conduction. GLA supplementation helps correct the impaired nerve function in animal models of diabetes. (Horrobin, *Diabetes*, 1997)

In the case of neuropathy and other neurological disorders, the concept of fatty acid balance needs to be considered. Deficiencies of Omega-3 fatty acids, combined with the inability to convert LA to GLA, influences membrane integrity, receptors, prostaglandin balance, and etc. Supplementation of a combination of GLA, EPA and DHA, plus antioxidant protection through alpha lipoic acid (300 mg. twice daily) can help correct symptoms of diabetic neuropathy. (Cameron, Horrobin)

Using Fat for Weight Control

The statement, "The fat you eat is the fat on your body", is actually only half true. Eating ***excess*** amounts of certain types of fat (refined and saturated) does lead to weight gain, but eating ***inadequate*** amounts of EFAs, can also lead to weight gain. Increasing your intake of two types of polyunsaturated fats can actually lead to weight loss.

The concept of ***"using fat as a fat-loss aid"*** may be a difficult one for many people to understand as we are constantly told to avoid fat. Udo Erasmus, in his book, *Fats that Heal, Fats that Kill,* reminds us of the following three important points concerning essential fatty acids:

1. EFAs increase our metabolic rate and energy production. Saturated fats slow down our metabolic rate making it easier to store excess calories.

2. Increased energy makes us feel better and makes us feel more active which can help increase muscle tone - which makes us look better. Increased muscle also helps us burn off more fat.

3. EFAs are required for the body to produce series 3 prostaglandins, which act as a diuretic to encourage elimination of excess water, and makes insulin work more efficiently in the body so we can burn glucose for energy.

We Eat Too Much, We Eat The Wrong Foods

Most people believe that the way to weight loss is to cut back on caloric intake and to increase physical activity. Consuming more calories than the body can burn off does lead to weight gain. However, cutting back on caloric intake can cause the body's metabolic rate to slow down making it more more difficult to lose weight. Reducing caloric intake to below 2,000 calories a day also makes it difficult to obtain adequate amounts of all essential nutrients, making it more difficult to burn calories and fat efficiently.

Excess calories are stored in the body as fat, but there is too much emphasis placed on the caloric value of foods. The real truth is that there are many, many reasons why we may put on excess weight...many more reasons than can be covered in a short essay.

Two of the major reasons for excess weight are poor food choices and malnutrition. We eat too many foods with low nutrient value (junk food), sometimes called empty calories. These foods have little or no nutritional value other than the calories they provide. If we cannot use these calories, they are readily stored as fat. Refined sugars and starches are readily converted into saturated fat and are stored by the body. These foods are the main cause of cellulite in the body.

Refined foods and starches often contain too many calories and too little fiber. They usually lack most of the nutrients required for efficient energy production and tremendously increase the body's requirement for them. Unrefined foods naturally contain these nutrients, but they are lost in processing.

When there is inadequate intake of all essential nutrients, fat is not efficiently burned. Fat is burned only if sufficient energy is produced. Energy production depends on almost every known nutrient.

For example, the B vitamins are important to produce energy. Fat is burned at a greatly reduced rate if vitamin B-5 and protein are under supplied. Vitamin B-6 is necessary for the energy conversion of stored fat. It is also a factor in the utilization of protein and fat. Proteins are needed for the proper function of many energy-producing enzymes. Protein cannot be used without other nutrients such as vitamin B-6 and choline.

Vitamin E is necessary for fat utilization. Sufficient amounts of E actually doubles our ability to use fat. Lecithin aids the cells to burn fat. If we are deficient in the nutrients necessary for production of lecithin, specifically choline and inositol, poor fat utilization results. Instead, the fat is stored.

Fat Calories Are Fatter!

Numerous clinical studies show that you put on more body fat by eating fat than by eating the same amount of calories from carbohydrates and protein. There are two reasons for this:

1. Evolution (survival of the fittest theory) has programmed humans to store fat against times of food shortages. This was necessary several centuries ago, but today we should be more concerned with eating too much rather than too little.

2. The use of body fat for energy is not related to fat intake. For energy, we first burn carbohydrates, then protein, and finally fat. It is rare that we actually burn the fats consumed. Most, not used for production of hormones, etc, goes directly to storage.

The only fat the body "uses" (not burns) are the EFAs, which are polyunsaturated. Unsaturated fats are much more easily used for energy by the body in comparison to saturated fats (usually from animal and dairy products). The body does not really use dietary saturated fats in any way but to store them - in your thighs, hips, rear, stomach, etc.

Although the body can convert excess carbohydrates and proteins to fat, it takes two grams of either to make one gram of fat. The metabolic conversion process uses about a quarter of the calories contained in the excess. So it's a lot of work for the body to convert complex carbohydrate and protein calories to stored fat. You really have to eat a lot of vegetables before you grow much fat.

Note: Fat cells cannot be changed into muscle cells, they must be burned off by muscle cells. Be very careful of statements made by diet plans or diet products promising to "change fat into muscle." This is impossible.

Avoid Saturated Fats!

Saturated fatty acids are found in animal products such as cheese, meat products, eggs, etc. Unless you are a strict vegetarian, who consumes no eggs or dairy products, it may surprise you to learn how much saturated fat is hiding in your diet. Here are some suggestions to help you cut down on saturated fat whenever possible and you'll be healthier and leaner in no time:

■ Reduce intake of red meat and pork. Choose leaner cuts of poultry and turkey without the skin.

■ Steam, broil, or bake foods instead of sauteing or frying.

■ Instead of butter, flavor potatoes with fresh pressed garlic, salsa, or herbs.

■ Instead of butter crackers, eat rye crisps, soda crackers, breadsticks - and hold the butter.

■ Instead of fattening salad dressing, use lemon juice or vinegar.

■ Instead of ice cream, substitute non-fat frozen yogurt, sherbet, sorbet, or ice milk.

■ Instead of butter or oil in baked goods, substitute an equal amount of apple sauce or plum sauce.

■ Thicken soups with pureed vegetables or beans instead of cream or corn starch.

Avoid Foods High in Saturated Fat: *% estimate is of total fat*

Coconut oil and **coconut butter** *(89% saturated)*
Palm and palm kernel oil *(46% saturated)*
Cottonseed oil *(23% saturated)*
Butter *(60% saturated)*
Hard cheese, i.e. cheddar *(63% saturated)*
Cream cheese *(64% saturated)*
Beef fat *(50% saturated)*
Pork/bacon *(37% saturated)*
Lamb chops *(54% saturated)*

Cholesterol: The Truth

Cholesterol has received a bad reputation in regards to heart health. It turns out that elevated cholesterol levels are not a reliable indicator of heart disease. It is true that cholesterol, because it is a saturated fat, is not exactly healthy. Saturated fat is readily stored and elevated weight does increase one's chances of developing heart disease, as well as many other degenerative conditions.

The body obtains cholesterol in two ways, through the diet and it is manufactured by the liver and intestines, but also by all cells in the body. We need cholesterol to make sex hormones, vitamin D, and bile salts, and regulate nerve and brain function.

No matter how much dietary cholesterol we eat, the liver will still go on making it. In fact, a diet high in saturated fats (such as cholesterol) and refined carbohydrates causes an increase in the body's production of cholesterol. Because the body produces it, there really is no need to consume it. Vegetarians do very well without it, and actually, have a much lower incidence of weight problems and heart disease.

Other Fats to Avoid

Hydrogenated and partially hydrogenated fats are fats which have been altered from their natural liquid form to become more solid (and stable). This processing alters the natural molecular structure of the fatty acids into an unnatural trans-configuration. These trans-fatty acids are unhealthy and should be avoided. They resemble saturated fats, but the body has a much more difficult time processing them. Trans-fatty acids also produce free radicals in the body. Trans-fatty acids are very uncommon in nature, and they are considered by the body to be "foreign."

Trans-fatty acids tend to raise LDL cholesterol levels and lower HDL cholesterol levels in the blood. In addition, these trans-fatty acids can easily become trapped along arterial walls creating an ideal environment for build-up of plaque and development of atherosclerosis.

In addition to creating trans-fatty acids, processing of fats destroys EFAs and other valuable nutrients such as carotenes, vitamin E and lecithin which help the body metabolize the fat. Examples of hydrogenated fat products include margarine, margarine-based products, shortening, and fats used for frying.

No-fat, low-fat and even trans-free tub margarines have recently become available, but it is too early to determine if these are completely healthy. The best solution may not involve *which* spread to use, but to use much less, or none at all, if possible.

Also be careful of "FAT FREE" products. Often the fat is replaced with refined sugar, which is very likely to turn into saturated fat in the body. Starches can also turn into saturated fats in the body so limit your consumption of potatoes, pasta, white rice, white bread, etc.

Increase Omega-3's EFAs

In his book, Dr. Erasmus references a women in California who actually lost 80 pounds of excess fat by adding three tablespoons of fresh flax seed oil to her diet. She had already been eating a nutritious diet for some time without weight loss success. This can be very frustrating and actually lead to setbacks and binge eating.

Studies done on obese rats showed that increasing their intake of Omega-3 fats with fish oil resulted in a decrease of body weight. There was also a decrease in LA delta-6 desaturation "bad guy" classes of prostaglandins and an increase in LNA delta-6 desaturation "good guy" classes of prostaglandins. (Ulmann)

Human studies have demonstrated that obese children have elevated levels of AA compared to healthy controls. They also showed higher levels of the Omega-6 long-chain polyunsaturated fatty acids, dihomo-gamma-linoleic acid (DGLA) and GLA than controls and plasma glucose levels were inversely related to AA. The researchers concluded that the higher levels of the long-chain Omega-6s in obese children was caused by an enhanced activity of delta-6-desaturation (due to an elevated intake of LA compared to LNA). They also speculated that the elevated insulin levels seen in obese individuals may stimulate production of LA delta-6 desaturation "bad guy" classes of prostaglandins. (Decsi) This can be avoided through supplementation of fish oils.

Other Helpful Fats

Conjugated Linoleic Acid, (CLA) is found in beef and dairy products. CLA is available in capsules derived from sunflowers. Studies on CLA suggest that it may help reduce body fat by directly affect fat metabolism. It may preferentially burn fat instead of glycogen during exercise and fasting. It also may counteract the adverse effect of cortisol. This is an adrenal hormone created by stress, such as severe illness and overtraining, which elevates sugar levels, breaks down muscle protein and increases fat deposition. CLA is also believed to increase production of series 1 prostaglandins which increases brain levels of somatotropin which increases growth hormone production. Recommended supplemental dosage for adults is between three and five grams daily.

GLA promotes cholesterol normalization and is a precursor to prostaglandin hormone compounds which stimulate energy production and calorie burning. Brown fat, which is believed to be the most metabolically active fat in the body, can be encouraged to burn up more fat with GLA. GLA has been shown to stimulate the brown fat and boost the body's use of its fat stores. Sources include cold-water fish, hemp, borage, black current and evening primrose.

Immune System

There is so much research in the area of EFAs, prostaglandins and leukotrienes that there is an entire medical journal published monthly devoted to the subject; *Prostaglandins, Leukotrienes and Essential Fatty Acid.*

The immune system is the internal defense system of the body designed to protect the body against disease. White blood cells, lymph cells, antibodies and interferon, are a few of the substances defending us from invaders. When something foreign enters the body, the immune system recognizes it and sends out signals to deal with it.

Immune disorders generally develop when the body's system of defense against foreign invaders goes out of control and starts attacking the body itself. Normally, the immune system is kept under control by the body's essential fatty acid-based regulatory system. Dietary imbalances, especially a shortage of the Omega-3 fatty acids, are known to cause or contribute to a breakdown of the immune system through prostaglandin imbalances and other abnormal reactions.

For a healthy, balanced immune system a balanced intake of Omega-3 and Omega-6 fatty acids is required. The reduced intake of Omega-3, combined with excessive Omega-6 intake, which is prevalent in the diets of most, leads to the overproduction of irritating Omega-6 prostaglandins.

The World Health Organization's recommendation for dietary intake of Omega-3 fatty acids is at least 3% of one's total caloric intake. Animal studies suggest that more than 3% is desirable for the regulation of eicosanoid metabolism. If one has a high intake of saturated fats (more than 10% of total calorie intake), the need for Omega-3 fatty acids increases. Without adequate Omega-

3s available, humans tend to synthesize more prostanoids, which is undesirable. If saturated fat intake is higher than that, the recommendation is 5-6% Omega -3s. (Dupont)

DHA Does Not Suppress Immune Functions

Studies conducted with purified EPA have shown that it may cause a decline in certain types of human immune response. Researchers have found that the addition of purified DHA to the diet did not inhibit many of the immune responses that have been reported by others to be inhibited when fish oils or EPA was added to the diets. This may be because DHA does not have the same effects as EPA. DHA supplementation has not shown any adverse health effects. (Kishida, Bechoua)

EPA and DHA Maintain Immune Function and Resistance to Bacterial Infection

The Omega-3 fatty acids from marine sources are known to have anti-inflammatory effects on monocyte function. There is some controversy whether Omega-3s may increase the susceptibility to infections. Researchers examining the effect of EPA and DHA on monocyte phagocytosis and respiratory burst activity found this to be untrue.

Fifty-eight healthy men took either a daily supplement of 3.8 g highly purified EPA, 3.6 g DHA, or corn oil for seven weeks. They found that monocytes retained their phagocytic (debris clean-up) ability and respiratory burst activity after supplementation. They found no reduction in internalization of bacteria and demonstrated that dietary EPA and DHA improved bacterial adherence to the monocyte surface. They concluded that monocytes retain their phagocytic potential after supplementation with purified EPA and DHA. (Halvorsen)

DHA Supports T and B Cells

Studies at the Department of Microbiology and

Immunology, Indiana University-Purdue University at Indianapolis,demonstrated additional beneficial effects on white blood cells with DHA. They observed that both T and B cell recoveries were increased by DHA. The effects were dose-dependent with a higher response rate for B cells. The researchers suggested that the DHA was enhancing the cellular membrane properties. They also suggested that the data implied that DHA may check ongoing immune response while concurrently preserving lymphocytes needed for subsequent immune responses. (Scherer)

One of the reasons to ensure Omega-3 fatty acid intake is adequate is because they are needed to regulate eicosanoid and prostaglandin production. AA enhances the tissue factor expression of mononuclear cells by the cyclo-oxygenase-1 pathway. Monocytes express tissue factor (TF) upon stimulation by inflammatory agents. Dietary administration EPA and DHA results in an impairment of TF expression by monocytes.

Researchers in France demonstrated that contrary to AA, EPA or DHA did not increase production of inflammatory agents (TXB2 or TF). The TF-dependent procoagulant activity was increased by AA; the combined action of PGG2 and TXA2, which potentiated it, was greater than that of PGE2, which inhibited it. (Cadroy)

Researchers at the School of Medicine, University of California, Davis, showed that diets rich in EPA and DHA can alter various macrophage functions. Specifically, their results showed that DHA and EPA can enhance the PAF-signaling pathway in macrophages by increasing the activation potential of phospholipase C, without affecting PAF receptor number and affinity. (Chakrabarti)

Cancer

The relationship between EFAs and cancer is complicated yet very simple at the same time. A healthy immune system is the best defense against all types of cancer. As cell division occurs daily in our body, mutations are a common occurrence which a healthy immune system quickly destroys before it can cause harm. When the immune system is unable to control these mutations, cancer results.

Many cancers, breast, colon and prostate, in particular, are correlated with a high fat intake. We also know that an excess intake of Omega-6 fats, even though they are essential, can increase cancer risk, increase tumor size, and increase the number of tumors.

In contrast, Omega-3 fats delay the formation of tumors, slows the rate of growth, can decrease tumor size and the number of tumors. Prostaglandins are a very important aspect of the immune system and largely regulated by our intake of fatty acids.

Eating Fish May Prevent Colon and Breast Cancer

In 1973 evidence was already emerging in population studies that suggested a link between total fat intake and certain cancers, most notably breast and colorectal cancer. The same studies suggested that consumption of animal fat was more closely related to cancer risk than fat intake overall.

In the last decade, further investigations into the links between diet and cancer have refined this equation. The most significant finding is that fish oil may act differently from other fats, protecting against these two cancers rather than promoting them, thus indicating a protective effect that may counteract some of the harm done by saturated fatty acids.

Studies that looked at different kinds of fat separately found that while all saturated fats and some polyunsaturated fats (such as the Omega-6 polyunsaturated fats found in margarine) were associated with cancer, monounsaturated fats (such as olive oil) and fish Omega-3 fats were inversely related to cancer risk. Animal studies have confirmed that while Omega-6 fats cause breast and colorectal tumors to grow, Omega-3 fats protect against them. (Caygill)

A team of British researchers examined data dating back to 1961 from 24 European countries to see if those with higher fish intake had lower rates of the two cancers, and vice versa. They found their hypothesis to be absolutely correct.

Country	Cancer rate per 100,000	Animal fat consumed	Fish consumed
Portugal	Lowest (colon) 7 (breast) 14	Lowest < 50 g/day	Low
Romania	Low	< 50 g/day	Low
Greece	(colon) 7.5 (breast) 15.2	51.9 g/day	48.2 g/day
Ireland	(colon) 21 (breast) 26.1	117.4 g/day	32.6 g/day
Iceland	Low	100 g/day	Highest 249 g/day

The relationship between animal fat and the two cancers was strong enough that the countries with the lowest intakes, such as Portugal, Greece, Spain, Romania and Bulgaria, without exception had lower rates of both cancers than the countries where intake was highest, such as the United Kingdom, Ireland, Belgium and Denmark.

However, the figures revealed a few apparent anomalies. Iceland was among the six or seven countries with over 100 g/day animal fat intake averaged over the whole period, yet its cancer risk was comparable to that of nations such as Portugal and Spain. Icelanders were marginally safer than the Portuguese, the lowest consumers of animal fat in all of Europe. The Icelanders also consume almost twice as much fish (91 kg every year) as the Europe's second-biggest seafood lovers, again the Portuguese. They consume four times as much as the Belgians (Europe's leading consumers of animal fat), 10 times more than the Dutch, and 20 times more than the Hungarians.

Another nation with higher than average animal fat consumption has even lower aggregate risk of breast and colorectal cancer than Iceland: Finland, Europe's third highest consumer of fish oil.

Collating all of this data, epidemiologists saw immediately that the role of animal fat in both cancers was far stronger than the role of fish consumption alone. That means that it is not a good idea to eat a lot of animal fat and try to compensate by eating a lot of fish. For example, Romanians, who eat relatively few fish, had Europe's lowest aggregate risk of these two cancers, because they ate less animal fat than almost anybody.

Fish oil consumed was found to be related to risk of both cancers, in both sexes, strongly evincing a protective tendency.

Researchers then divided the 24 nations into high animal fat (over 85 g/day) and low animal fat groups. Fish consumption was found to mark a very striking difference in risk among high-fat countries, but to make little difference in those countries where little animal fat is consumed.

In general, fish, fish oil, and animal fat intake had a stronger impact on colorectal cancer risk than on breast cancer risk, and stronger protective effect in men than in women.

The researchers made the interesting point that while the 15% reduction in animal fat intake recommended by Britain's Department of Health would reduce the colorectal cancer death rate among men by only about 6%, the drop could be increased to about 30% if combined with supplementation of 1,000 mg of fish oil a day.

EFA Slow Abnormal Cellular Proliferation

Abnormal cellular proliferation is associated with cancer cells which tend to be faster growing than normal cells. Studies show that EPA, DHA and GLA significantly reduce this dangerous (cytotoxic) activity without affecting healthy cells. EPA and DHA seem to have stronger protective abilities than GLA. (Purasiri)

DHA Suppress Cancer Growth

Norwegian researchers found that in the case of DHA, this appears to be inaccurate. They exposed lung cancer cell lines to DHA, discovering that its presence was important to prevent proliferation of the cancer when stimulated. They explained that DHA supported the fatty acid composition of the membranes. (Schonberg)

Researchers in Japan investigated the effects of EPA and DHA on human breast cancer cells and demonstrated that both EPA and DHA suppress breast cancer cell proliferation. (Noguchi)

DHA Found to Inhibit Leukemia

Japanese researchers examined human eosinophilic leukemia (Eol-1) cells for their ability to generate platelet-activating factor (PAF) and the effect of supplementation of DHA on PAF synthesis. The fatty acid composition of phospholipids and AA were also evaluated.

Although undifferentiated cells did not produce PAF, the exposure of IFN-gamma differentiated leukemia cells to generate PAF in response to the Ca-ionophore. In addition, the IFN-gamma-treated cells acquired the ability to release free fatty acids, approximately 55% of which were AA.

When DHA was supplemented into the leukemia cell culture for 24 hours, PAF production decreased by 40 to 50%. Supplementation of EPA did not significantly decrease PAF production. (Shikano)

Brain Cancer Tissue Deficient in DHA

Researchers in the U.S. have found that healthy brain cell tissue differs in the fatty acid composition from brain cancer cells. They compared the fatty acid composition of tumor tissue from glioma (brain cancer) patients with that of normal brain tissue. They found that levels of DHA were significantly reduced in the glioma samples compared with normal brain samples. This reduction in glioma DHA content was also observed in terms of phospholipids (4.6 vs. 9.6). The phosphatidylserine and phosphatidylethanolamine were also reduced in the glioma samples.

Differences were also noted in the Omega-6 content between glioma and normal brain samples. The glioma content of the LA was significantly greater than that observed in the control samples in terms of total lipids. (Martin)

DHA, EPA, LNA Suppress Nitric Oxide Production

Nitric oxide (NO) is an important biological mediator, but it is also a dangerous free radical and its excessive production in inflammation is thought to be a causative factor for cellular injury and, over the long term, cancer. Researchers at the National Cancer Center Research Institute in Tokyo, demonstrated that Omega-3 fatty acids suppressed nitric oxide production in macrophage cells. Suppression of nitric oxide production was observed with DHA, EPA and LNA in a dose-dependent fashion. In contrast, no inhibition was observed with LA, the Omega-9 PUFA (oleic acid) or a saturated fatty acid (stearic acid). The inhibitory effect of these Omega-3s on nitric oxide production in activated macrophages could contribute to their cancer chemopreventive influence. (Ohata)

Another research team at the Department of Internal Medicine, University of Texas, Southwestern Medical School, Dallas, also showed that DHA inhibits production of nitric oxide (NO) by macrophages. (Khair-El-Din T)

DHA Useful for Chemotherapy

Canadian researchers have discovered in animal studies that supplementation of DHA enhanced bone marrow growth and substantially increased macrophage white blood cells, suggesting expansion of the bone marrow compartment with DHA feeding. This suggests that DHA supplementation may be useful in conjunction with chemotherapy. (Atkinson) Chemotherapy, which is designed to kill "fast growing" cells, not only kills cancerous cells, but can also destroy healthy blood cells. If one's white blood cell count becomes too low, chemotherapy must be discontinued.

Other research of interest...

DHA levels depleted in infertile men

The morpho-functional indexes of ejaculates from 35 healthy and 145 infertile men were studied. The depression of functional sperm activity was found in the majority of infertile men. Phospholipid analysis of whole ejaculates from 12 healthy and 35 infertile subjects was performed.

It was shown that compared to healthy men, infertile men had decreased levels of total phospholipids (including Lyso-phosphatidyl choline and phosphatidyl inositol) and DHA.

The quantity of DHA in ejaculates of infertile men was positively correlated with motility of spermatozoa. (Hula)

Salmon Recipes

Salmon is easy to cook and elegant to serve. Here are a few of my favorites.

Roasted Salmon

Prep. time: 10 minutes (plus1 hour marinating).
Cooking time: about 10 minutes. Serves 4.

4 6-8oz Salmon Fillets, center-cut, skin on
1 Tbs. olive oil
1 Tbs. fresh lemon juice
1 Tbs. chopped fresh rosemary
Salt and freshly ground pepper, to taste

Dry salmon with paper towel. Whisk together the oil, lemon juice, rosemary and pepper. Rub onto salmon, covering all sides, place on a plate, cover loosely and allow to marinate refrigerated for 1 hour.

Preheat oven to 425° F (220° C). If your non-stick skillet does not have an oven-proof handle double wrap it in aluminum foil.

Brush skillet with olive oil and preheat.

Remove salmon from fridge and lightly season with salt.

Place salmon flesh side down in very hot pan to sear for 1 minute. Immediately place hot pan with salmon into the hot oven and roast for 8 minutes.

Salmon should flake when pressed with a fork when done.

Grilled Salmon

Use whole salmon of any type. Leave the skin on as it lifts right off after cooking. Place the salmon on a piece of foil large enough to completely wrap it. Fill the inside of the salmon with a mixture of chopped onion, celery, carrots, garlic and green peppers. Add 2 or 3 slices of lemon. Add salt and pepper to taste if desired. Seal the foil and cook on a grill or in an oven. The cooking time depends on the size/amount of fish. Open the foil and test the fish for flakeyness. The vegetables usually absorb a lot of fat/oil from the fish.

Easy Grilled Salmon

from the kitchen of Beth and Randy

Fresh salmon fillet of any type with skin on. (We like sock-eye) The fillet size is according to number of people (or dal-mations) you are serving.

Brush flesh side of fillet with olive oil and salt lightly.

Place flesh side down onto hot grill for a few minutes and turn 1/4 turn to obtain "X" grill marks. Cook for a few minutes longer and carefully turn over. I like to cover the fish when it is cooking skin side down to prevent any moisture loss. The fish is done when it flakes apart when pressed with a fork. The skin peels off easily.

(This also works well for steelhead trout and other firm fish)

Poached Salmon

Small salmon fillets, approximately 6 ounces, are poached by putting about a half inch of water in a small, 5-6 inch fry pan, covering it, heating the water to simmer, then putting in the fillet covered for four minutes. Add whatever seasoning you like to the salmon or to the water. The four minutes leaves the center uncooked and very juicy. Cook longer if desired.

If there's skin, it usually sticks to the pan, and if you run hot water over it as soon as you remove the salmon, it'll wash/scrape right off.

Fillets may be served whole or cut it into inch and a half wide pieces and added to a dark green salad with tomato, ripe avocado, red onion, croutons, and any tasty dressing.

Citrus Salmon

from Gerd Voigt of the Winchelsea House Restaurant in Lantzville, BC.

Prep time: 10 minutes/Cooking Time: 7-10 minutes

1 lb. salmon fillets
Salt and pepper
1 Tbs. cornstarch
1 Tbs. water
2 Tbs. undiluted frozen orange juice concentrate
1 Tbs. lemon juice
1/4 cup brown sugar
Garnish (optional): 1 sliced orange/parsley

Sprinkle both sides of the salmon fillet with salt and pepper. Mix the cornstarch and water in a small bowl to form a paste. Add the orange juice concentrate, lemon juice and brown sugar. Stir mixture well until all the ingredients are dissolved. Set aside.

Pour half of the sauce into the bottom of a microwaveable dish. Place the salmon fillet in the dish on top of sauce. Pour the remaining sauce over the salmon. Cover the dish with plastic wrap. Vent to allow steam to escape.

Microwave on high for 7-10 minutes (depending on microwave).

Remove from microwave and remove plastic wrap. Place the fillet on a plate.

Stir remaining sauce and pour over the fillet and garnish if desired.

Fusilli with Baked Tomatoes and Salmon

From "Totally Salmon" by Helene Siegel, Celestial Arts

1/8 cup olive oil
5 medium tomatoes, cored and cut in small wedges
1 lb. skinless salmon fillet, cut in 1 inch cubes
1/2 cup thinly sliced fresh basil or serrel
1 Tbs. minced garlic

salt and freshly ground pepper
1 lb. fusilli or corkscrew pasta

Preheat oven to 400 degrees F.

Lightly coat a medium ceramic or glass baking dish with some of the oil. Cover the bottom half with half of the tomatoes. Top with half of the salmon, the garlic, and basil or sorrel. Salt and pepper. Repeat the layers and seasoning. Drizzle lightly with rest of oil.

Bake uncovered for 20 to 25 minutes until tomatoes soften and juices run together, stirring once.

While baking, cook the pasta in a large pot of salted water, drain and transfer to a serving bowl. Top with roasted sauce and toss well. Serves 4.

Salmon-Broccoli Loaf with Dill and Capers

from Jean Anderson's Processor Cooking

Very moist, let it stand at room temp 30 minutes before slicing. Recipe needs no salt...

1 cup loosely packed parsley sprigs, washed and patted dry on paper towels
6 slices firm-textured white bread
2 cups 1/2 inch cubes of broccoli stems
1 medium yellow onion, cut into slim wedges
1 3/4 pounds cooked or canned boned salmon (remove all dark skin)
1/3 cup drained capers (use the small capers)
2/3 cup light cream (I use non-fat evaporated milk)
4 eggs
2 Tbs. snipped fresh dill or 3/4 tsp dill weed
Finely grated rind of 1/2 lemon
1/8 tsp. freshly ground black pepper

In food processor fitted with metal chopping blade, mince parsley fine, using 5-6 on-offs of the motor; empty into a large mixing bowl. Crumb the bread 2 slices at a time, with two or three 5-6 on-offs; add to bowl. Dump all the broccoli stems into processor; mince very fine with about three 5-second bursts; add to bowl.

In the processor-mince the onion -- 3 or 4 bursts -- add to the bowl.

Flake the salmon in three batches -- 2 on-offs will be enough. Add to the mixing bowl along with all remaining ingredients. Mix thoroughly, pack mixture firmly into a well-buttered 9x5x3 inch loaf pan and bake in a slow oven (300 degrees F.) for about 1 hour and 40 minutes, or until loaf begins to pull from sides of pan and is firm to the touch. Remove loaf from oven and let it stand upright in its pan on a wire rack for 30 minutes. Carefully loosen the loaf all around with a thin-bladed spatula, then invert gently onto a large serving platter.

Salmon Limone

One of my favorites adapted from Spasso's Family Italian Restaurant in Laguna Hills, CA

3/4 lb. skinless salmon fillet, poached/cut into two pieces
1/3 cup olive oil
1/2 cup lemon juice
2 Tbs. capers
Minced garlic to taste
Bite-sized pieces of carrot, cauliflower, and broccoli -lightly steamed
Cooked angle hair pasta

Mix olive oil and lemon juice and combine with vegetables and capers, toss lightly in pasta, divide onto plates and place salmon on top. Serves 2.

Smoked Salmon with White Bean Soup

A tasty idea I discovered at King's Fishhouse in Laguna Hills, CA. Locations also in Long Beach and Orange, CA.

Serves 16 cups or 8 bowls
3 lbs. dry navy beans
3 lbs. pre-cooked hot-smoked salmon cut into 1/2 inch cubes
1 1/4 cups diced celery

1 1/4 cups diced carrots
1 1/4 cups diced onion
1/2 bunch chopped cilantro
4 cloves chopped garlic
4 ounces butter or olive oil
4 ounces flour
1 quart tomato sauce
2 Tbs. fresh thyme (or 1 tbls. dried)
2 bay leaves
1 Tbs. paprika
2 Tbs. black pepper
2 cups clam juice
2 quarts water

Cook beans separately in unsalted water until al dente. Drain water and set aside.

In large heavy soup kettle cook celery, carrots, onion, cilantro and garlic in butter or olive oil until vegetables are transparent. Add flour and cook until mixture starts to brown and smells nutty.

Add remaining ingredients except for salmon and cook about 20 minutes (until beans are tender) on medium to low heat. Add salmon. Serve when salmon is warmed through. Perfect with fresh sourdough bread (and a glass of cabernet).

Spicy Salmon and Eggplant

Recipe by: Judy Dinardo-Namak

3 fresh salmon steaks
1 eggplant
3 limes
2 lemons
1/8 cup olive oil
1 tsp. Greek or Italian seasoning
1/8 tsp. hot red pepper flakes
Fresh ground black pepper

Remove stem and end of eggplant and slice on a diagonal, cutting slices approximately 1/4-inch thick.

Place salmon steaks and eggplant slices in a large, flat

plastic container.

Cut lemons and limes in half and remove juice. Pour juice into a separate bowl. Stir in olive oil, seasoning, pepper flakes and ground pepper. Pour over steaks and eggplant. Cover and marinate in refrigerator for 1 to 1-1/2 hours.

Turn steaks over and rearrange eggplant for even marinating, once during process.

Cook steaks on a hot grill on both sides until done.

Add eggplant slices to grill when fish is half cooked. Grill eggplant slices on both sides. Remove.

Serve fish and vegetables with rice. Serve hot.

Thai Salmon Parcels for 2

from Delia Smith's summer collection

2 4-5 ounce salmon fillets
4 sheets filo pastry
1 ounce butter
Zest & juice of 1 lime
1 tsp. grated ginger
1 clove garlic (pressed)
1 spring onion (finely chopped)
1 Tbs. fresh coriander (finely chopped)
Salt & pepper

Mix together lime zest and juice, garlic, spring onion, ginger and coriander.

Melt butter. Lay out 1 sheet of filo, and brush with butter. Lay second sheet on top, brush with more butter. Lay a salmon fillet about 2-3 inches from short side of pastry, season to taste and put half of lime mixture on top.

Fold short end of pastry over salmon, then fold in the 2 long sides. Fold the salmon over twice more, and cut off the remaining pastry. Do the same with the other fillet.

Put the parcels on a well-greased baking sheet, and just before baking brush with the remaining butter. Cook at medium heat for 20-25 minutes, until brown and crispy.

Glossary

Alpha Linolenic Acid (LNA): An Omega-3 long-chain polyunsaturated fatty acid containing 18 carbons and 3 double bonds, (18:3n-3). It cannot be manufactured in the body and is therefore an essential part of the daily diet.

Arachidonic acid (AA): An Omega-6 long-chain polyunsaturated fatty acid containing 20 carbons and 4 double bonds, (20:6n-4). It is not essential in adults as it can be derived from LA. It is found in animal fats and is often consumed in excess in modern diets.

Cholesterol: A saturated fat found in animal products. It is not essential as it is produced in the body and used to form steriod hormones. Excess dietary cholesterol is considered detrimental.

Delta-6-desaturase: Enzyme needed to convert linoleic acid (LA) to gamma linolenic acid (GLA) and alpha linolenic acid (LNA) to steridonic acid (SDA) which converts to EPA and then DHA.

Docosahexaenoic Acid (DHA): An Omega-3 polyunsaturated fatty acid derived from LNA containing 22 carbons and six double bonds (22;6n-6). It contains the highest number of double bonds of all fatty acids and therefore is most susceptible to damage.

Docosapentaenoic acid (DPA): An Omega-6 polyunsaturated fatty acid derived from LA containing 22 carbons and five double bonds (22;5n-6). It is sometimes used if there are adequate supplies of DHA.

Eicosapentaenoic Acid (EPA:) An Omega-3 polyunsaturated fatty acid derived from LNA containing 20 carbons and five double bonds (20;5n-6). It is used to produce anti-inflammatory protaglandins, PGE3.

Elongation: Enzymatic process by which a fatty acid is lengthened by 2 carbon atoms.

Gamma Linolenic Acid (GLA): An Omega-6 long -chain polyunsaturated fat. Used to increase the body's production of the anti-inflammatory prostaglandin PGE1. Also recommended when the enzyme delta-6 desaturase is dysfunctional in the body.

Lecithin: Important emulsifier needed to break down fats for digestion. Found in soy and egg yolk. Also a source of phosphatidylcholine and phosphatidyserine.

Linoleic Acid (LA): An Omega-3 long-chain polyunsaturated fatty acid containing 18 carbons, 2 double bonds, (18:2n-6). It is an essential part of the daily diet, although most diets contain an over-abundance.

Lipid: Another name for fat or fatty acid.

Lipid peroxidase: Rancid (or damaged) fats caused by free radical damage to an unsaturated fatty acid. This causes damage to cell membranes.

Monounsaturated fats: Although not essential, these unsaturated fatty acids containing one double bond are generally considered to be among the healthier vegetable fats; found in olive and sunflower oil.

Phosphatidylcholine: Important component of nerve cell membranes made up of two fatty acids, phosphate and choline. Found in lecithin.

Phosphatidylinositol: Important component of nerve cell membranes made up of two fatty acids, phosphate and inositol.

Phosphatidyserine: Important component of nerve cell membranes made up of two fatty acids, phosphate and serine. Also found in lecithin.

Polyunsaturated fatty acid: A fatty acid that contains more than one double bond between carbon atoms in its chain. They include the Omega-3 polyunsaturated fats found in flax, fish and fish oil, and the Omega-6 polyunsaturated fats, such as those found in margarine or safflower oil.

Saturated fats: Fatty acids containing no double bonds. Generally considered to be the least desirable fats in their effect on human metabolism, cancer and cardiovascular risk; found in animal and dairy products, and in some nuts.

Steridonic acid (SDA): An Omega-3 polyunsaturated fatty acid directly derived from LNA containing 18 carbons and four double bonds (18;4n-3).

Transfatty acid: An unsaturated fatty acid that has been altered from its natural curved shape to an unnatural crooked shape. This occurs through hydrogenation of vegetable oils (normally fluid at room temerature) to produce margarine - in attempts to make it solid at room temperature - like butter. The body treats these as saturated fats, they should be avoided.

Unsaturated fats: Fatty acids that contain multiple double bonds. The higher the number of double bonds the more sensitive it is to damage.

Bibliography

Adachi J; Tatsuno Y; Imamichi H; Ninomiya I; et al Identification of cholesta-3,5-dien-7-one by gas chromatography-mass spectrometry in the erythrocyte membrane of alcoholic patients.Alcohol Clin Exp Res 1996 Feb;20(1 Suppl):51A-55A.

Adams, P. et al; Arachidonic Acid to EPA ratio in blood coorelates positively with clinical symptoms of depression. Lipids 1996; Suppl.) 31:S-157-61.

Agren JJ; et al,; Fatty acid composition of erythrocytes, platlet, and serum lipids in strict vegetarians. Lipids 1995;30(4):365-69.

Agren JJ; Vaisanen S; Hanninen O; Muller AD; Hornstra G; Hemostatic factors and platelet aggregation after a fish-enriched diet or fish oil or docosahexaenoic acid supplementation. Department of Physiology, University of Kuopio, Finland. Prostaglandins Leukot Essential Fatty Acids 1997 Oct;57(4-5):419-21.

Agostoni C; Trojan S; Bellu R; Riva E; et al; Developmental quotient at 24 months and fatty acid composition of diet in early infancy: a follow up study. Department of Paediatrics, University of Milan, San Paolo Hospital, Italy. Arch Dis Child 1997 May;76(5):421-4.

Arm JP; Horton CE; Spur BW; Mencia-Huerta JM; Lee TH; The effects of dietary supplementation with fish oil lipids on the airways response to inhaled allergen in bronchial asthma. Am Rev Respir Dis 1989 Jun;139(6):1395-400.

Arm JP; Thien FC; Lee TH; Leukotrienes, fish-oil, and asthma. Department of Rheumatology, Brigham and Women's Hospital, Boston, Massachusetts Allergy Proc 1994 May-Jun;15(3):129-34.

Atkinson TG; Meckling-Gill KA; Barker HJ; Incorporation of long-chain n-3 fatty acids in tissues and enhanced bone marrow cellularity with docosahexaenoic acid feeding in post-weanling Fischer 344 rats. Lipids 1997 Mar;32(3):293-302.

Bechoua S; Dubois M; Nemoz G; Lagarde M; Prigent AF; Docosahexaenoic acid lowers phosphatidate level in human activated lymphocytes despite phospholipase D activation. J Lipid Res 1998 Apr;39(4):873-83.

Blusztajn JK; Liscovitch M; Mauron C; Richardson UI; Wurtman RJ. Phosphatidylcholine as a precursor of choline for acetylcholine synthesis. Department of Brain and Cognitive Sciences, Massachusetts Institute of Technology, Cambridge. J Neural Transm Suppl 1987;24:247-59.

Boehm G; Muller H; Kohn G; Moro G; Minoli I; Bohles HJ. Docosahexaenoic and arachidonic acid absorption in preterm infants fed LCP-free or LCP-supplemented formula in comparison to infants fed fortified breast milk. Ann Nutr Metab 1997;41(4):235-41.

Broughton KS; Whelan J; Hardardottir I; Kinsella JE; Effect of increasing the dietary (n-3) to (n-6) polyunsaturated fatty acid ratio on murine liver and peritoneal cell fatty acids and eicosanoid formation. Institute of Food Science, Cornell University, Ithaca, NY 14853. J Nutr 1991 Feb;121(2):155-64.

Broughton KS; Johnson CS; Pace BK; Liebman M; Kleppinger KM Reduced asthma symptoms with n-3 fatty acid ingestion are related to 5-series leukotrieneproduction. Am J Clin Nutr 1997 Apr;65(4):1011-7.

Cameron NE; Cotter MA; Horrobin DH; Tritschler HJ. Effects of alpha-lipoic acid on neurovascular function in diabetic rats: interaction with essential fatty acids. UK. Diabetologia 1998 Apr;41(4):390-9.

Carlson SE. Arachidonic acid status of human infants: influence of gestational age at birth and diets with very long chain n-3 and n-6 fatty acids. J Nutr 1996 Apr;126(4 Suppl):1092S-8S.

Carlstrom K; Eriksson S; Rannevik G Sex steroids and steroid binding proteins in female alcoholic liver disease. Acta Endocrinol (Copenh) 1986 Jan;111(1):75-9

Carnielli VP; Wattimena DJ; Luijendijk IH; Boerlage A; et al; The very low birth weight premature infant is capable of synthesizing arachidonic and docosahexaenoic acids from linoleic and linolenic acids. Pediatr Res 1996 Jul;40(1):169-74.

Caygill C, Charlett A, Hill M; Fat, fish, fish oil and cancer. Br J Cancer 1996; 74(1):159-64.

Cadroy Y; Boneu B; Dupouy D; Arachidonic acid enhances the tissue factor expression of mononuclear cells by the cyclo-oxygenase-1 pathway: beneficial effect of n-3 fatty acids. J Immunol 1998 Jun 15;160(12):6145-50.

Chakrabarti R; Erickson KL; Lim D; Hubbard NE; Alteration of platelet-activating factor-induced signal transduction in macrophages by n-3 fatty acids. University of California, Davis Cell Immunol 1997 Jan 10;175(1):76-84.

Chaudry AA; Moffat LE; McClinton S; Wahle KW; Arachidonic acid metabolism in benign and malignant prostatic tissue in vitro: effects of fatty acids and cyclooxygenase inhibitors. Int J Cancer 1994 Apr 15;57(2):176-80.

Christensen O; Christensen E; Fat consumption and schizophrenia. Copenhagen Chest Clinic, Denmark. Acta Psychiatr Scand 1988 Nov;78(5):587-91

Clandinin MT; Van Aerde JE; Parrott A; et al; Assessment of the efficacious dose of arachidonic and docosahexaenoic acids in preterm infant formulas: fatty acid composition of erythrocyte membrane lipids. Pediatr Res 1997 Dec;42(6):819-25

Cockburn, F;, Neonatal brain and dietary lipids, Archieves of Disease in Childhood, 1994, 70 F1-F2.

Corrigan FM; Van Rhijn AG; Horrobin, DF.; EFAs in Alxheimer's Disease. Ann NY AcadAci , 1991;6640:250-2.

Corrigan FM; Mowat B; Skinner ER; Van Rhijn AG; Cousland G; High density lipoprotein fatty acids in dementia. Argyll and Bute NHS Trust, Argyll and Bute Hospital, Lochgilphead, UK. Prostaglandins Leukot Essential Fatty Acids 1998 Feb;58(2):125-7.

Crawford, M.; The role of essential fatty acids in neural development: implications for perinaral nutrition. Am Journal of Clin Nutr, 1993: 57 (suppl): 703S-710S.

Crook, T., et al; Effects of phosphatidlyserine in Alzheimer's disease. Psychopharmacol Bull, 1992;28:61-6.

Davidson MH; Maki KC; Kalkowski J; Schaefer EJ; Torri SA; Drennan KB Effects of docosahexaenoic acid on serum lipoproteins in patients with combined hyperlipidemia: a randomized, double-blind, placebo-controlled trial. J Am Coll Nutr 1997 Jun;16(3):236-43

De Caterina R; Libby P; Gimbrone MA Jr; Clinton SK Cybulsky MA Omega-3 fatty acids and endothelial leukocyte adhesion molecules. Harvard Medical School, Prostaglandins Leukot Essential Fatty Acids 1995 Feb-Mar;52(2-3):191-5

Decsi T; Molnar D; Koletzko B Long-chain polyunsaturated fatty acids in plasma lipids of obese children. Lipids 1996 Mar;31(3):305-11

DiGiacomo RA; Kremer JM; Shah DM Fish-oil dietary supplementation in patients with Raynaud's phenomenon: a double-blind, controlled, prospective study. Division of Rheumatology, Albany Medical College, New York 12208. Am J Med 1989 Feb;86(2):158-64

Dry J; Vincent D Effect of a fish oil diet on asthma: results of a 1-year double-blind study. Centre d'Allergie, Hopital Rothschild, Paris, France. Int Arch Allergy Appl Immunol 1991;95(2-3):156-7

Dupont J Essential fatty acids and prostaglandins. Prev Med 1987 Jul;16(4): 485-92

Edwards R; Peet M; Shay J; Horrobin D Omega-3 polyunsaturated fatty acid levels in the diet and in red blood cell membranes of depressed patients. J Affect Disord 1998 Mar;48(2-3):149-55

Fasching P; Ratheiser K; Waldhausl W; Rohac M; Osterrode W; Nowotny P;Vierhapper H Metabolic effects of fish-oil supplementation in patients with impaired glucose tolerance. Diabetes 1991 May;40(5):583-9

Fergusson, DM, Beautrais, AL, Silva, PA, Breast-feeding and cognitive development in the first seven years of life. Soc Sci. Med 1982, Vol 16, pp 1705-1708

Foreman-van Drongelen, et al, Long-chain polyunsaturated fatty acids in preterm infant: Status at birth and its influence on postnatal levals., Jour of Ped, April 1995.

Friedberg CE; Janssen MJ; Heine RJ; Grobbee DEFish oil and glycemic control in diabetes. A meta-analysis. Department of Internal Medicine, Ziekenhuis der Vrije Universiteit, Amsterdam, The Netherlands. Diabetes Care 1998 Apr;21(4):494-500

Gatti P; Viani P; Cervato G; Testolin G; Simonetti P; Cestaro B Effects of alcohol abuse: studies on human erythrocyte susceptibility to lipid peroxidation. Facolta di Medicina, Universita degli Studi Milano, Italy. Biochem Mol Biol Int 1993 Aug;30(5):807-17.

Gibson RA; Effect of increasing breast milk docosahexaenoic acid on plasma and erythrocyte phospholipid fatty acids and neural indices of exclusively breast fed infants. Eur J Clin Nutr 1997 Sep;51(9):578-84.

Glen AI; Horrobin DF; Brayshaw N; Vaddadi K; Rybakowski J; Cooper SJ; Membrane fatty acids, niacin flushing and clinical parameters. Prostaglandins Leukot Essential Fatty Acids 1996 Aug;55(1-2):9-15.

Glen AI; Skinner FS; Ellis K; Morse-Fisher N ; Vaddadi KS; Horrobin DF; A red cell membrane abnormality in a subgroup of schizophrenic patients: evidence for two diseases Schizophr Res 1994 Apr;12(1):53-61.

Grimsgaard S; Bonaa KH; Hansen JB; Nordoy A Highly purified eicosapentaenoic acid and docosahexaenoic acid in humans have similar triacylglycerol-lowering effects but divergent effects on serum fatty acids. Institute of Community Medicine, University of Tromso, Norway. Am J Clin Nutr 1997 Sep;66(3):649-59.

Glueck CJ; Bates SR Migraine in children: association with primary and familial dyslipoproteinemias. Pediatrics 1986 Mar;77(3):316-21.

Haban P; Simoncic R; Klvanova I; Ozdin L; Zidekova E [The effect of n-3 fatty acid administration on selected indicators of cardiovasculardisease risk in patients with type 2 diabetes mellitus] Klinicke oddelenie Vyskumneho ustavu vyzivy v Bratislave. Bratisl Lek Listy 1998 Jan;99(1):37-42.

Halvorsen DS; Hansen JB; Grimsgaard S; et al The effect of highly purified eicosapentaenoic and docosahexaenoic acids on monocyte phagocytosis in man. University of Tromso, Norway. Lipids 1997 Sep;32(9):935-42.

Hamazaki T; Sawazaki S; Itomura M; Asaoka E; Nagao Y; et al The effect of docosahexaenoic acid on aggression in young adults. A placebo-controlled double-blind study. J Clin Invest 1996 Feb 15;97(4):1129-33.

Hamazaki T; et al, DHA emulsion of blood glucose and insulin contrations in diabetic rats. Annals of the new York Academy of Sciences, Vol. 683, 1993, 207-212.

Hansen JB; Grimsgaard S; Nilsen H; et al Effects of highly purified eicosapentaenoic acid and docosahexaenoic acid on fatty acid absorption, incorporation into serum phospholipids and postprandial triglyceridemia. Lipids 1998 Feb;33(2):131-8.

Heird WC; Prager TC; Anderson RE; Docosahexaenoic acid and the development and function of the infant retina. Department of Pediatrics, Baylor College of Medicine, Houston, Texas, USA. Curr Opin Lipidol 1997 Feb;8(1):12-6.

Hibbeln, J., Salem, N.; Dietary polyunsaturated fatty acids and depression. Am J Clinical Nutrition 1995; 62, 1-9.

Hoffman, DR, Birch, DG.; DHA in red blood cells of patients with X-linked retinitis pigmentosa. Invest Ophalmol Vis Sci 1995;36(6):1009-18.

Holbrook PG; Wurtman RJ; Calcium-dependent incorporation of choline into phosphatidylcholine (PC) by base-exchange in rat brain membranes occurs preferentially with phospholipid substrates containing docosahexaenoic acid (22:6(n-3)), Massachusetts Institute of Technology, Cambridge. Biochim Biophys Acta 1990 Sep 18;1046(2):185-8.

Holman, R., et al; A case of human kinikenic acid deficiency involving neurological abnormalities. Am J Clin Nutr 1982;35:617-23.

Horrobin DF; Essential fatty acids in the management of impaired nerve function in diabetes. Scotia Research Institute, Kentville, Nova Scotia, Canada. Diabetes 1997 Sep;46 Suppl 2:S90-3.

Horrobin DF; Schizophrenia as a membrane lipid disorder which is expressed throughout the body. Scotia Pharmaceuticals, Stirling, UK. Prostaglandins Leukot Essential Fatty Acids 1996 Aug;55(1-2):3-7.

Horrobin DF; The relationship between schizophrenia and essential fatty acid and eicosanoid metabolism. Efamol Research Institute, Kentville, Nova Scotia, Canada. Prostaglandins Leukot Essential Fatty Acids 1992 May;46(1):71-7.

Horrobin DF; Fatty acid metabolism in health and disease: the role of delta-6-desaturase. Efamol Research Institute, Kentville, Nova Scotia, Canada. Am J Clin Nutr 1993 May;57(5 Suppl):732S-736S; discussion 736S-737S.

Horrobin DF The effects of gamma-linolenic acid on breast pain and diabetic neuropathy: possible non-eicosanoid mechanisms. Efamol Research Institute, Kentville, Nova Scotia, Canada. Prostaglandins Leukot Essential Fatty Acids 1993 Jan;48(1):101-4.

Horrobin DF Scotia Research Institute, Kentville, Nova Scotia Essential fatty acids in the management of impaired nerve function in diabetes. Canada. Diabetes 1997 Sep;46 Suppl 2:S90-3.

Horwood, LJ; Fergusson, DM; Breastfeeding and Later Cognitive and Academic Outcomes Pediatrics 1998 Jan, Vol 101, pg. e.

Horby Jorgensen M; Holmer G; Lund P; et al; Effect of formula supplemented with docosahexaenoic acid and gamma-linolenic acid on fatty acid status and visual acuity in term infants. J Pediatr Gastroenterol Nutr 1998 Apr;26(4) 412-21.

Howe, P.; Can we recommend fish oil for hypertension? Clin. Exp. Pharmacol. Physiol. 1995, 22: 199-203.

Howe P.; Dietary Fats and Hypertension, Lipids and Syndromes of Insulin Resistance Annals of the NYAS: 1997, Vol 827, pg 339-352.

Hudson CJ; Warsh JJ; Cashman F; Cogan S; Lin A; The niacin challenge test: clinical manifestation of altered transmembrane signal transduction in schizophrenia? Clarke Institute of Psychiatry, University of Toronto, Ontario,Canada. Biol Psychiatry 1997 Mar 1;41(5):507-13.

Hula NM; Tron'ko MD; Volkov HL; Marhitych VM; [Lipid composition and fertility of human ejaculate] Ukr Biokhim Zh 1993 Jul-Aug;65(4):64-70.

Kelley, D, Taylor P, Nelson G, Mackey B; Dietary DHA and Immunocompetence in Humans, United States Department of Agriculture National Agricultural Library.

Khair-El-Din T; Lu CY; Miller RT; et al; Transcription of the murine iNOS gene is inhibited by docosahexaenoic acid, a major constituent of fetal and neonatal sera as well as fish oils. J Exp Med 1996 Mar 1;183(3):1241-6.

Khalfoun B; Lebranchu Y; Bardos P; Thibault G; Docosahexaenoic and eicosapentaenoic acids inhibit in vitro human lymphocyte-endothelial cell adhesion. Transplantation 1996 Dec 15;62(11):1649-57.

Kinsella, J. et al; Dietary n-3 polyunsaturated fatty acids and amelioration of cardiovascular disease;possible mechanisms. Am. J. Clin. Nutr. 1990, 52:1-28.

Kishida E; Yano M; Kasahara M; Masuzawa Y; Distinctive inhibitory activity of docosahexaenoic acid against sphingosine-induced apoptosis. Hyogo University of Teacher Education,Yashiro, Hyogo 673-14, Japan. Biochim Biophys Acta 1998 Apr 22;1391(3):401-8.

Kjeldsen-Kragh J; Lund JA; Riise T; Finnanger B; et al; Dietary omega-3 fatty acid supplementation and naproxen treatment in patients with rheumatoid arthritis. J Rheumatol 1992 Oct;19(10):1531-6.

Klein A; Bruser B; Malkin A The effect of fatty acids on the vulnerability of lymphocytes to cortisol. Department of Clinical Biochemistry, Sunnybrook Medical Centre, University of Toronto, Canada. Metabolism 1989 Mar;38(3):278-81.

Kragballe Dietary supplementation with a combination of n-3 and n-6 fatty acids (super gamma-oil marine) improves psoriasis. Acta Derm Venereol 1989;69(3):265-8

Lassus A; Dahlgren AL; Halpern MJ; et al Effects of dietary supplementation with polyunsaturated ethyl ester lipids (Angiosan) in patients with psoriasis and psoriatic arthritis. Department of Dermatology, University Central Hospital, Helsinki, Finland. J Int Med Res 1990 Jan-Feb;18(1):68-73.

Lau CS; Morley KD; Belch JJ Effects of fish oil supplementation on non-steroidal anti-inflammatory drug requirement in patients with mild rheumatoid arthritis--a double-blind placebo controlled study. Department of Medicine, Ninewells Hospital, Dundee, Scotland. Br J Rheumatol 1993 Nov;32(11):982-9.

Leaf AA; Leighfield MJ; Costeloe KL; Crawford MAFactors affecting long-chain polyunsaturated fatty acid composition of plasma choline phosphoglycerides in preterm infants. Joint Medical College of St. Bartholomew's Hospital, London, England. J Pediatr Gastroenterol Nutr 1992 Apr;14(3):300-8.

Lee TH; Arm JP; Horton CE; Crea AE; Mencia-Huerta JM; Spur BW; Effects of dietary fish oil lipids on allergic and inflammatory diseases. Department of Allergy and Allied Respiratory Disorders, U.M.D.S., Guy's Hospital, London, U.K. Allergy Proc 1991 Sep-Oct;12(5):299-303.

Maehle L; Haugen A; Krokan HE; Schonberg S; Mollerup S; Effects of n-3 fatty acids during neoplastic progression and comparison of in vitro and in vivo sensitivity of two human tumor cell lines. Department of Toxicology, National Institute of Occupational Health, Oslo, Norway. Br J Cancer 1995 Apr;71(4):691-6

Maire JC; Wurtman RJ; Choline production from choline-containing phospholipids: a hypothetical role in Alzheimer's disease and aging. Prog Neuropsychopharmacol Biol Psychiatry 1984;8(4-6):637-42.

Makrides M; Neumann M; Simmer K; Pater J; Gibson R; Are long-chain polyunsaturated fatty acids essential nutrients in infancy? Lancet: 1995 Jun 10 Issue/Part/Supplement: 345 Volume: 8963, 1463-8.

Mahadik SP; Scheffer RE; Correnti EE; Jenkins K, Horrobin DF; Plasma membrane phospholipid fatty acid composition of cultured skin fibroblasts from schizophrenic patients: comparison with bipolar patients and normal subjects. Psychiatry Res 1996 Jul 31;63(2-3):133-42.

Martin DD; Hussey DH; Wen BC; Spector AA; Robbins ME; The fatty acid composition of human gliomas differs from that found in nonmalignant brain tissue. Lipids 1996 Dec;31(12):1283-8.

McCreadie, R.; The Nithsdale Schizophrenia Syrveys 16. Breast-feeding and schizophrenia: preliminary results and hypotheses. 334-337.

McKeone BJ; Osmundsen K; Brauchi D; et al; Alterations in serum phosphatidylcholine fatty acyl species by eicosapentaenoic and docosahexaenoic ethyl esters in patients with severe hypertriglyceridemia. J Lipid Res 1997 Mar;38(3):429-36.

Melchert HU; Limsathayourat N; et al; Fatty acid patterns in triglycerides, diglycerides, free fatty acids, cholesteryl esters and phosphatidylcholine in serum from vegetarians and non-vegetarians. Atherosclerosis 1987 May;65(1-2):159-66.

Meydani SN; Lichtenstein AH; Cornwall S; Meydani M; Goldin BR; Immunologic effects of national cholesterol education panel step-2 diets with and without fish-derived N-3 fatty acid enrichment. USDA, Human Nutrition Research Center on Aging, Tufts University, Boston, J Clin Invest 1993 Jul;92(1):105-13.

Mitchell EA; Aman MG; Turbott SH; Manku M; Clinical characteristics and serum essential fatty acid levels in hyperactive children. Clin Pediatr (Phila) 1987 Aug;26(8):406-11.

Mollerup S; Haugen A; Differential effect of polyunsaturated fatty acids on cell proliferation during human epithelial in vitro carcinogenesis: involvement of epidermal growth factor receptor tyrosine kinase. Br J Cancer 1996 Aug;74(4):613-8.

Monteleone P; Beinat L; Tanzillo C; Maj M; Effects of phosphatidylserine on the neuroendocrine response to physical stress in humans. Neuroendocrinology 1990 Sep;52 (3): 243-8.

Nelson GJ; Schmidt PC; Bartolini GL; Kelley DS; Kyle D; The effect of dietary docosahexaenoic acid on plasma lipoproteins and tissue fatty acid composition in humans. USDA, San Francisco, California Lipids 1997 Nov;32(11):1137-46.

Neuringer, M.; Reisbick, S.; Janowsky, J.; The Role of n-3 fatty acids is visual and cognitive development: Current edidence and methods of assesment. The Journal of Pediatrics, 1994, Vol. 125,:5, Part 2 S39-S47.

Newman PE; Could diet be one of the causal factors of Alzheimer's disease? Source: Med Hypotheses, 1992 Oct, 39:2, 123-6.

Ng LL; Davies JE; Quinn PA; Ngkeekwong F; Uptake mechanisms for ascorbate and dehydroascorbate in lymphoblasts from diabetic nephropathy and hypertensive patients. Leicester Royal Infirmary, UK. Diabetologia 1998 Apr;41(4):435-42.

Nightengale, S., et al.; Red blood cell and adipose tissue fatty acids in active and inactive mutiple sclerosis. Acta Neurol Scand 1990; 82: 43-50.

Noguchi M; Earashi M; Minami M; Kinoshita K; Miyazaki I Effects of eicosapentaenoic and docosahexaenoic acid on cell growth and prostaglandin E and leukotriene B production by a human breast cancer cell line (MDA-MB-231). Oncology 1995 Nov-Dec;52(6):458-64.

Noguchi M; Miyazaki I; Earashi M; Rose DP; The role of fatty acids and eicosanoid synthesis inhibitors in breast carcinoma. Kanazawa University Hospital, School of Medicine, Japan. Oncology 1995 Jul-Aug;52(4):265-71.

Ohata T; Wakabayashi K; Sugimura T; Takahashi M Fukuda K; Suppression of nitric oxide production in lipopolysaccharide-stimulated macrophage cells by Omega-3 polyunsaturated fatty acids. J Cancer Res 1997 Mar;88(3):234-7.

Pawlosky RJ; Salem N Jr Ethanol exposure causes a decrease in docosahexaenoic acid and an increase in docosapentaenoic acid in feline brains and retinas. National Institute on Alcoholism and Alcohol Abuse, National Institutes of Health, Rockville, MD 20852, Am J Clin Nutr 1995 Jun;61(6):1284-9.

Peet M; Horrobin D; Shay J; Murphy B Depletion of omega-3 fatty acid levels in red blood cell membranes of depressive patients. Biol Psychiatry 1998 Mar 1;43(5):315-9.

Peet M; Ramchand CN; Mellor J; L Essential fatty acid deficiency in erythrocyte membranes from chronic schizophrenic patients, and the clinical effects of dietary supplementation. Prostaglandins Leukot Essential Fatty Acids 1996 Aug;55(1-2):71-5.

Pita ML; Delgado MJ; Carreras O; et al Chronic alcoholism decreases polyunsaturated fatty acid levels in human plasma, erythrocytes, and platelets--influence of chronic liver disease. Thromb Haemost 1997 Aug;78(2):808-12.

Prasad MR; Markesbery WR; Dhillon H; Yatin MLovell MA Regional membrane phospholipid alterations in Alzheimer's disease. Department of Surgery, University of Kentucky, Lexington Neurochem Res 1998 Jan;23(1):81-8.

Raederstorff D; Moser U; Bachmann H; Pantze Anti-inflammatory properties of docosahexaenoic and eicosapentaenoic acids in phorbol-ester-induced mouse ear inflammation. Allergy Immunol 1996 Nov;111(3):284-90.

Reddy, S, Sanders, TAB, et al,; The influence of maternal vegetarian diet on the essential fatty acid status of the newborn. Uero J Clin Nutr 1994;48;358-68.

Rhodes LE; White SI; Dietary fish oil as a photoprotective agent in hydroa vacciniforme. Department of Dermatology, Royal Liverpool University Hospital, U.K. Br J Dermatol 1998 Jan;138(1):173-8.

Roberts LJ 2nd; Morrow JD; Dettbarn WD; Formation of isoprostane-like compounds (neuroprostanes) in vivo from docosahexaenoic acid. Vanderbilt University, Nashville, Tennessee J Biol Chem 1998 May 29;273(22):13605-12.

Rodriguez, B. et al.; Fish Intake May Limit The Increase In Risk Of Coronary Heart Disease Morbidity And Mortality Among Heavy Smokers The Honolulu Heart Program. Circulation. Vol. 94, 1996:952-956.

Salem N Jr, et al. The nervous system has an absolute molecular species requirement for proper function. Mol Membr Biol. 1995 Jan; 12(1): 131-134.

Salem N Jr; Pawlosky RJ; Docosahexaenoic acid is an essential nutrient in the nervous system. Laboratory of Membrane Biochemistry and Biophysics, DICBR, National Institute of Alcohol Abuse and Alcoholism, Bethesda, MD 20892. J Nutr Sci Vitaminol (Tokyo) 1992;Spec No:153-6.

Sanders, T., Reddy, S; The infludence of a vegetarian diet on the fatty acid composition of milk and the EFA satus of the infant. J Pediatr 1992; 120:S71-7.

Scherer JM; Jenski LJ; Stillwell W; Spleen cell survival and proliferation are differentially altered by docosahexaenoic acid. Purdue University at Indianapolis Cell Immunol 1997 Sep 15;180(2):153-61.

Scheurlen M; Kirchner M; et al; Fish oil preparations rich in docosahexaenoic acid modify platelet responsiveness to prostaglandin-endoperoxide/ thromboxane A2 receptor agonists. Universitat Tubingen, Germany. Biochem Pharmacol 1993 Jul 20;46 (2):245-9.

Schonberg SA; Skorpen F; Paracetamol counteracts docosahexaenoic acid-induced growth inhibition of A-427 lung carcinoma cells and enhances tumor cell proliferation in vitro. Anticancer Res 1997 Jul-Aug;17(4A):2443-8.

Shikano M; Yazawa K; Masuzawa Y; Effect of docosahexaenoic acid on the generation of platelet-activating factor by eosinophilic leukemia cells, Eol-1. J Immunol 1993 Apr 15;150(8 Pt 1):3525-33.

Shoda R; Umeda N; Yamato S; Matsueda K; Therapeutic efficacy of N-3 polyunsaturated fatty acid in experimental Crohn's disease. J Gastroenterol 1995 Nov;30 Suppl 8:98-101.

Soyland E; Lea T; Sandstad B; Drevon A Dietary supplementation with very long-chain n-3 fatty acids in man decreases expression of the interleukin-2 receptor (CD25) on mitogen-stimulated lymphocytes from patients with inflammatory skin diseases. Eur J Clin Invest 1994 Apr;24(4):236-42.

Sperling RI; Benincaso AI; Knoell CT; Larkin JK; et al; Dietary omega-3 polyunsaturated fatty acids inhibit phosphoinositide formation and chemotaxis in neutrophils. Harvard Medical School, J Clin Invest 1993 Feb;91(2):651-60.

Stenson, W, et al; Dietary supplementation with fish oil in ulcerative colitis. Annals of Internal Medicine116:609-614, 1992.

Stevens LJ, Zentall SS, Deck J, et al; Essential fatty acid metabolism in boys with attention-deficit hyperactivity disorder. Purdue University, Am J Clin Nutr 1995 Oct;62(4):761-8.

Stordy BJ; Benefit of docosahexaenoic acid supplements to dark adaptation in dyslexics Lancet 1995 Aug 5;346(8971):385.

Stubbs, CD; The structure and function of DHA in membranes. Essential Fatty Acids and Eicosanoids. Champaign, Illinois: American Oil Chemist'Society, 1992: 116.

Swann PG; Parent CA; Croset M; Fonlupt P; Lagarde M; et al; Enrichment of platelet phospholipids with eicosapentaenoic acid and docosahexaenoic acid inhibits thromboxane A2/prostaglandin H2 receptor binding and function. J Biol Chem 1990 Dec 15; 265(35): 21692-7.

Taylor, B, Wadsorth, J; Breastfeeding and child development at five years Developmental Medicine and Child Neurology, 1984, 26. pg. 73-80.

Ulmann L; Blond JP; Poisson JP; Bezard J; Incorporation of delta 6- and delta 5-desaturation fatty acids in liver microsomal lipid classes of obese Zucker rats fed n - 6 or n - 3 fatty acids. Biochim Biophys Acta 1994 Aug 25;1214(1):73-8.

van der Tempel H; Tulleken JE; Limburg PC; Muskiet FA; van Rijswijk MH; Effects of fish oil supplementation in rheumatoid arthritis, State University Groningen, The Netherlands. Ann Rheum Dis 1990 Feb;49(2):76-80.

Virkkunen ME; Horrobin DF; Jenkins DK; Plasma phospholipid essential fatty acids and prostaglandins in alcoholic, habitually violent, and impulsive offenders. Forensic Psychiatric Department, Helsinki University Central Hospital, Finland. Biol Psychiatry 1987 Sep;22(9):1087-96.

Whelan J; Broughton KS; Lokesh B; Kinsella JE; In vivo formation of leukotriene E5 by murine peritoneal cells. Institute of Food Science, Cornell University, Ithaca, NY 14853. Prostaglandins 1991 Jan;41(1):29-42.

Whelan J; Broughton KS; Kinsella JE; The comparative effects of dietary alpha-linolenic acid and fish oil on 4- and 5-series leukotriene formation in vivo. Cornell University, Ithaca, New York 14853. Lipids 1991 Feb;26(2):119-26.

Williams LL; Kiecolt-Glaser JK; Horrocks LA; Quantitative association between altered plasma esterified omega-6 fatty acid proportions and psychological stress. Prostaglandins Leukot Essential Fatty Acids 1992 Oct;47(2):165-70.

Wilkinson DI; Do dietary supplements of fish oils improve psoriasis? Psoriasis Research Institute, Palo Alto, California 94301. Cutis 1990 Oct;46(4):334-6.

Yazawa, K.; Clinical esperience with DHA in demented patients, International Conference on Highly Unsaturatd Fatty Acids in Nutritio and Desise Prevention, Barcelone, Spain, 1996, November 4-6.

Yu G; Bjorksten B; Serum levels of phospholipid fatty acids in mothers and their babies in relation to allergic disease. Eur J Pediatr 1998 Apr;157(4):298-303.

Index

ADD, 5, 10, 19, 41-43, 101-103, 105-107
Addiction, ... 3, 5, 43, 47
ADHD, ...5, 19, 40
Aging, ... 21, 55, 59, 88
Albacore Tuna, ... 9
Alcohol, ...11, 21, 27, 44, 46-47
Allergies, ...3, 5, 21, 41, 58, 62
Alpha Lipoic Acid, ...21, 49, 77, 83
Alpha-Linolenic Acid, ... 7-8, 25, 32
see also Omega-3
Alzheimer's Disease, 3, 5, 19, 38, 55-57
Anti-inflammatory Drugs, ...69
Antioxidants 21, 38, 49, 56-57, 77-78, 81, 83
Arachidonic Acid, 8, 25-26, 28, 31-32, 46
Arthritis, 3, 5, 11, 26, 28, 31, 58-59, 65-70, 83
Aspirin, ...11, 21, 28, 58
Asthma, ...3, 5, 31, 41, 58-59, 62-64
B Cells, ...92-93
B-3, ...30, 52
B-6, ...30, 41
Behavior, 3, 5, 19, 24, 37, 41, 43, 46, 48, 58
Brain, 3, 5-6, 15, 17, 19-24, 26-27, 33-34, 37-39, 49-51, 53, 55-57, 88, 90, 98
Brain Cancer, ...98
Brain Growth, ...33-34
Breast Cancer, ...94, 96-97
Breast-fed, ... 19-20, 33-34, 37
Brown fat .90
Cancer, ...3, 5, 26, 31, 58, 77, 94-99
Cellulite 85
Cerebral Cortex, ...22, 37
Cerebral Palsy, ...39
Chemotherapy, ...99
Cholesterol, 5, 26, 45, 48-49, 59-60, 69, 73-75, 88-90
Cigarettes, ... 21, 77-78
Cod, ...15-16
Conjugated linoleic acid 90
Corn, ...8-10, 61, 87, 92
Coronary Artery Disease, ... 3, 73, 80
Cortisol, ...43-44, 47, 90
Crohn's Disease 3, 72
Delta-6-desaturase 7, 20, 29, 90
Dementia, ... 19, 21, 38, 55-56
Depression, 3, 5, 19, 27, 47-50, 56, 100
Dermatitis, ...5, 19, 31, 58-59, 65
DGLA, ...14, 25, 40, 44-45
DHEA, ...6
DHGLA, ...62
Diabetes, ...3, 5, 20, 27, 80-83
Docosatetraenoic acid (DEA), ...46
Down Syndrome, ...38
Docosapenaenoic acid (DPA) 22
Dyslexia, ...19, 24, 53
Eczema, ...19, 21, 26, 41, 58
Edema, ...13, 61, 63
Eggplant, ...105-106
Eggs, ...9-10, 16-17, 32, 87, 104
Eicosapentaenoic Acid, ... 7-8, 14, 25
Eskimos, ...73
Essential Fatty Acids, 3, 7, 9-11, 25, 35, 38, 44, 49, 56, 62, 84
Evening Primrose Oil, ...9, 30, 90
Fish, 4, 6, 8-9, 11, 14-20, 30, 32, 36, 46, 48-49, 54, 59-61, 63-67, 69-70, 72, 77-79, 81-82, 90, 92, 94-97, 102, 106
Fish Farming 11, 13
Flax, ...9, 32, 90
Gamma-Linolenic Acid, ... 25, 32, 90
Glucose, ...80-82, 84
Hair, ...12
HDL-cholesterol, ...74
Hemp, .9, 90,
HIV, ...83
Hydrogenized oils 11
Hypertension 77-78
IL-1, ...60-61
IL-4, ...71
IL-6, ...60-61, 71
IL-8, ...71
Immune System, ... 3, 26, 91, 94
Inflammation, 13, 26, 28-29, 31, 58, 61, 63, 68, 71-72, 98
Insulin, ...7, 20, 26, 80-81, 84, 89
IQ, ...20, 33-35, 37
LA, 7-8, 11-12, 16-17, 20, 25, 28-30, 32, 36-37, 44, 46, 83, 98
LDL-cholesterol, ...74
Lecithin, ...86, 89
Leucotriene B5, ...69
Leukotriene B4, ...61
Leukotrienes, 28, 31, 58-59, 63, 68, 70, 91
Linoleic Acid, 7-8, 11-12, 16-17, 20, 25, 27, 30, 32, 36-37, 44, 46, 83, 90, 98
Lipid, 11, 22, 35, 38, 47, 52, 55, 63, 73, 75, 81-82

Liver, .12, 15-16, 21, 27, 46, 75-76, 88
LNA, .8, 11-14, 16-17, 20, 24-25, 29-30, 33-34, 36-37, 59, 81, 83, 98
Memory, ...3, 5, 19, 21, 34, 55
Mental, ...13, 33-34, 45-46, 51, 55
Microalgae, ...8, 15
Miscarriage, ...12
Mitochondria, ...23
Monocytes, ...68, 71, 92-93
Monounsaturated, ...60, 95
Multiple Sclerosis, ...27, 49-50
Nerve damage 82-83
Nerves, ...22-24, 26, 35, 57, 86
Neurological, ... 19, 34-35, 83
Neuropathy, ... 82-83
Neutrophil, ...59, 68-69
Neurotransmitters 22-23. 57
Niacin, ...41, 51-53
NIDDM, ...81-82
Nitric Oxide, ...98-99
NSAID, ...70
Numbness, ...19, 67
Nuts, ...8, 32
Oleic Acid, ...8, 44, 98
Olive oil, 9-10, 64, 67, 95, 101-102, 105-107
Omega-3, 6-9, 11-12, 14-15, 17-18, 20, 25, 30, 34-36, 38, 41, 47-50, 54, 59-62, 64-65, 67-69, 73, 76, 81, 83, 91-95, 98
Omega-6, 7-9, 11-12, 14, 22, 25, 38, 41, 48-50, 59-60, 62, 69, 76, 81, 91, 94-95, 98
PAF, ...58, 63, 93, 97-98
PGE1, ...26-28, 30, 32, 45, 58
PGE2, 27-28, 31-32, 45, 58, 60-61, 76, 93
PGE3, ...20, 28, 30, 32
PGEl, ...28
PGG2, ...93
Phosphatidylcholine, ...44
Phosphatidyserine, ...44
Phospholipid, 44-45, 53, 57, 62, 69, 75-76, 83, 100, 117
Photoreceptors, ...23
Phototesting, ...66
Platelet, ...27-28, 31, 76-77, 79
PMS, ...27
Polymorphonuclear, ...61
Pregnancy, ...35, 39
Prostacyclin, ...31
Prostaglandin E2, ...31, 59, 61
Prostaglandin Regulation, ...29
Prostaglandins, 11, 15, 18, 25-26, 28-32, 41-42, 44, 53, 58-59, 61-62, 68, 70, 72-73, 76-77, 81, 83-84, 90-91, 93-94,
Prostate, ...27, 94
Psoriasis, ...5, 19, 31, 58-59, 65-66
Rancid fats 7, 10, 38, 55
REM, ...43
Retinitis, ...21, 24
Safflower, ... 8-9
Salmon, , . 9, 15-17, 32, 101-107, 118-119
Salmon Recipes, , ... 101
Saturated fats .7, 10-11, 17, 20, 27, 30, 42, 48, 52, 54, 59-60, 84-89, 91-92, 94-95, 98
Schizophrenia, ...3, 19, 26, 51-54
Seafood, ...16-17, 54, 96
Sjogren Syndrom, ...27
Skin, 3, 12-13, 19, 41, 46, 60-61, 65, 83, 87, 101-102, 104
Skin Disorders, ...3, 19, 65
SOD, ... 38
Soybean, ...9, 45
Sterility, ...12
Sunflower, ...9-10
Synaptic membrane, ... 22-23, 49, 56
T-cells, ...26, 65
Tardive Dyskinesia, ...54
Trans-fatty acids, ...42, 88
Tromboxane A2, ...31
Trout, ...9, 15, 32
Tuna, ...9, 16-17, 32, 116
Ulcerative Colitis, ... 72
UV Radiation, ... 66
Vegetable Oils, ...8, 30
Vegetarians, ...18-19, 88
Violence, ...5, 19
Violent, ...44-45
Vision, ... 13, 19, 22-24
Visual Development, ...35
Vitamin A, ...16
Vitamin C, ...21, 41
Zinc, ...30, 41

About The Author...

Beth M. Ley Jacobs has been a science writer specializing in health and nutrition for over 10 years. She wrote her own undergraduate degree program and graduated in Scientific and Technical Writing from North Dakota State University in 1987 (combination of Zoology and Journalism). Beth has her masters and doctoral degrees in Nutrition.

Beth lives in southern California with her husband, Randy, and two dalmations, Cameo and K.C. She is dedicated to spreading the health message, exercises on a regular basis, eats a vegetarian, low-fat diet and takes anti-aging supplements.

Memberships: American Academy of Anti-aging, New York Academy of Sciences, Oxygen Society.

About The Cover Artist...

Roderick Sutterby lives and works in Northumberland, England. He studied painting and ceramics at Bath Academy of Art, Corsham, and has been a full-time wildlife and landscape painter since 1995. His emphasis is on conservation and the environment with a particular interest in the wild salmon. He is preparing an illustrated book on the creatures and their life struggles.

A segment of his oil painting "The Valley" is featured on the cover. More of his work can be seen at the website, http://www.imu.ac.uk/ces/axis/ or at http:/www.riverdale.k12.or.us/salmon/art/sutterby/index.html.
email: sutterby@globalnet.co.uk